T0135762

Bibliographic information published by the Deutsche Nationalbibliothek

The Deutsche Nationalbibliothek lists this publication in the Deutsche Nationalbibliografie; detailed bibliographic data are available in the Internet at http://dnb.d-nb.de .



ISBN 978-3-8325-3621-3

Logos Verlag Berlin GmbH
Comeniushof, Gubener Str. 47,
10243 Berlin
Tel.: +49 (0)30 42 85 10 90
Fax: +49 (0)30 42 85 10 92
INTERNET: http://www.logos-verlag.de

# Electronic Microwave Imaging with Planar Multistatic Arrays

Faculty of Engineering
Friedrich-Alexander-Universität
Erlangen-Nürnberg

for the doctoral degree

Doctor of Engineering
(Dr.-Ing.)

submitted by

**Sherif Sayed Aboelyazeed Ahmed**

Thesis accepted by
Faculty of Engineering
Friedrich-Alexander-Universität Erlangen-Nürnberg

| | |
|---|---|
| Oral examination date : | April 26, 2013 |
| Chair of the bodies responsible : | Prof. Dr.-Ing. habil. Marion Merklein |
| Reviewers : | Prof. Dr.-Ing. Lorenz-Peter Schmidt |
| | Prof. Dr.-Ing. Thomas Zwick |

# Electronic Microwave Imaging with Planar Multistatic Arrays

Der Technischen Fakultät
der Friedrich-Alexander-Universität
Erlangen-Nürnberg

zur Erlangung des Doktorgrades

Doktor-Ingenieur
(Dr.-Ing.)

vorgelegt von

**Sherif Sayed Aboelyazeed Ahmed**

Als Dissertation genehmigt von
der Technischen Fakultät
der Friedrich-Alexander-Universität Erlangen-Nürnberg

Tag der mündlichen Prüfung : 26. April 2013

Vorsitzender des Promotionsorgans : Prof. Dr.-Ing. habil. Marion Merklein

Gutachter : Prof. Dr.-Ing. Lorenz-Peter Schmidt
Prof. Dr.-Ing. Thomas Zwick

# Preface

Working in the domain of microwave imaging is a fortune, an opportunity for an interdisciplinary research connecting various scientific fields from electromagnetism far to image processing techniques. Having the chance to experience the birth of a highly demanded technology and to pioneer its shape, gives a special taste of delightfulness to each research step despite all difficulties one has to face. Driving the microwave imaging technology to reach a fully electronic realization requires a novel way in understanding the options made by the electromagnetic theories in the light of the modern digital era. Moreover, I have been very fortunate to be granted a possibility to work in an academic atmosphere, which is healthily bonded to industrial facilities. Thus scientifically sophisticated and technologically challenging solutions became within reach. Bridging different realms was indeed more of a pleasure than a task to accomplish.

On the road to all achievements, there is always a great companionship, persons who enlighten the way and give their hand. Among them, I would like to thank Prof. Dr.-Ing. Lorenz-Peter Schmidt for his great trust and remarkable support over the years. Special thanks are owed to Christian Evers, who continuously supported the research work even in the unfortunate moments. Further thanks go to Prof. Dr.-Ing. Thomas Zwick for his valuable comments and time. I am also grateful to Andreas Schießl, Dr.-Ing. Ralf Jünemann, Dr. Michael Aigle, Prof. Michael Hiebel, Claus Tremmel, Sebastian Methfessel, and Dr.-Ing. Frank Gumbmann for their enthusiasm and competence. A dedicated thank and appreciation are deserved to Dr.-Ing. Olaf Ostwald for the fruitful exchanges and his help in preparing the German summary. Space would not be sufficient to name all ambitious contributors to the demonstration of this work from Rohde & Schwarz in Munich as well as the Institute of Microwaves and Photonics in Erlangen. It was a great time collaborating with both. Last but not least, a heartily appreciation is deserved to my mother, without whom, it would not have been for this work to see the light.

I deeply hope that the following pages will add a significant contribution to the knowledge of the mankind and assist in advancing their capabilities towards a peaceful and prosperous life.

Munich, 2013

*Thanks Allah*

*For a better world...*
*I know you waited for this.*

# Contents

Page

Chapter 1

# Introduction

## 1.1 Light over time

The research on imaging systems and methods has started in the history of humanity as early as the first investigations on the human visual system, namely the eye. It is the human most important sensing organ which can intercept light to construct images of objects at largely varying distances. Therefore, the early investigations were made in the visible spectrum, where light is easily detected by the naked eye. The interests had ranged from the foundation of mirrors to the searching of methods to assist and enhance the visual capability of the human, like magnifying optics. From a scientific point of view, the inherit operation principle of the human eye was of a major interest, as well as the nature of light itself.

The ancient Greeks in Alexandria, located on the north shore of Egypt, conducted research on the properties of light. Euclid (estimated to have lived around 300 BC) discussed his observations on the visibility of a seen object in dependence on angular width. Heron of Alexandria (lived around AD 40) explained many ways for the usage of planar and curved mirrors, followed a century later by the work of Ptolemy (lived around AD 140) on reflection and refraction of light. Thanks to these efforts, early understanding of the nature of light was achieved. It was, however, believed that the eye emits rays of infinite velocity and receives them back to see an image. This thinking is interestingly clearly demonstrated by the words of Heron in his "Catoptrica". As the original work is lost, the text became available after several translations [1]. He told:

> *"That the rays proceeding from our eyes move with infinite velocity may be gathered from the following consideration. For when, after our eyes have been closed, we open them and look up at the sky, no interval of time is required for the visual rays to reach the sky. Indeed, we see the stars as soon as we look up, though the distance is, as we may say, infinite. Again, if this distance were greater the result would be the same, so that, clearly, the rays are emitted with infinite velocity. Therefore they will suffer neither interruption, nor curvature, not breaking, but will move along the shortest path, a straight line."*

It took, however, around a thousand years until the scientific work of Al-Hasan Ibn Al-Haytham, in Arabic الحَسن بن الهيثم and often known as Alhazen, could explain the human eye to be a passive receiver. His work was published around AD 1021 in Kitab al-Manazir, its original Arabic name is كتَاب المنَاظر and is known as Book of Optics [2]. In his original Arabic text, first chapter, in the starting of the 6th section, he wrote:

وَ قَد تبين أَيضَاً أن من خَاصة الضوء أن يُؤَثر في البصر[§] و أن من طبيعة البصر أن يَنفَعِل بِالضوء. فأَخلق بأَن يكون إِحسَاس البَصر بِالضوء الذي في المُبصر إِنمَا هو من الضوء الذي يرد منه إِلَى البصر.

This translates into: *And it has also been shown that it is the characteristic of the light to affect the eye and it is the nature of the eye to get influenced by the light. Therefore, it is appropriate that the sensing of the eye to the light, which is in the seen object, is actually caused by the light coming from the seen object to the eye.*

With these simple words, Ibn al-Haytham settled a new understanding of the human visual ability. He conducted also further research to analyze the operation of the human eye, along with much effort to explain the process of image focusing inside it based on the concept of rays. This effort in turn led to the first explanation of the formation of images using the camera obscura method known many centuries before [3].

Targeting a quantitative mathematical description for the nature of light, a long debate between scientists in the 17th and 18th centuries took place. Newton (1642 – 1727) was convinced of the corpuscular theory of light, where the mechanical image of the world was too dominant. He studied phenomena related to colors and the decomposition of white light using prisms. Nevertheless, his theory possibly hindered the development of the wave theory of light. His famous experiment leading to the so called Newton's rings was an evident for the interference phenomenon of light, which was, however, fully rejected by him; and instead he delivered explanations based on his corpuscular theory, as in "Opticks" [4] in 1704.

Contrarily to the corpuscular theory, description of light as a wave was considered by Huygens. Christiaan Huygens (1629 – 1695) presented a wave concept of the light, which was to some extent imitating the wave behavior recognized by sound waves. The light was thus assumed to propagate in spherical waves, for which each surface portion is considered a source for a secondary wave. With such, he could explain reflection and refraction phenomena. His foundations based on a conceptual similarity

[§]The word البصر in this old Arabic manuscript was meant to be the human eye.

with sound waves can be readily deduced from his book "Traité de la lumière" published in 1690. In its original French text [5] he explained:

> *"Nous savons que par le moyen de l'air, qui est un corps invisible et impalpable, le son s'étend tout à l'entour du lieu où il a été produit, par un mouvement qui passe successivement d'une partie de l'air à l'autre, et que l'extension de ce mouvement se faisant également vite de tous, côtés, il se doit former comme des surfaces sphériques qui s'élargissent toujours et qui viennent frapper notre oreille. Or il n'y a point de doute que la lumière ne parvienne aussi depuis le corps lumineux jusqu'à nous par quelque mouvement imprimé à la matière qui est entre deux, puisque nous avons déjà vu que ce ne peut être par le transport d'un corps qui passerait de l'un à l'autre. Que si avec cela la lumière emploie du temps à son passage, ce que nous allons examiner maintenant, il s'ensuivra que ce mouvement imprimé à la matière est successif et que par conséquent il s'étend, ainsi que celui du son, par des surfaces et des ondes sphériques : car je les appelle ondes, à la ressemblance de celles que l'on voit se former dans l'eau quand on y jette une pierre, qui représentent une telle extension successive en rond, quoique provenant d'une autre cause et seulement dans une surface plane."*

With its translation into English [6] is:

> *"We know that by means of the air, which is an invisible and impalpable body, Sound spreads around the spot where it has been produced, by a movement which is passed on successively from one part of the air to another; and that the spreading of this movement, taking place equally rapidly on all sides, ought to form spherical surfaces ever enlarging and which strike our ears. Now there is no doubt at all that light also comes from the luminous body to our eyes by some movement impressed on the matter which is between the two; since, as we have already seen, it cannot be by the transport of a body which passes from one to the other. If, in addition, light takes time for its passage which we are now going to examine it will follow that this movement, impressed on the intervening matter, is successive; and consequently it spreads, as Sound does, by spherical surfaces and waves: for I call them waves from their resemblance to those which are seen to be formed in water when a stone is thrown into it, and which present a successive spreading as circles, though these arise from another cause, and are only in a flat surface."*

Experimental verification was made later by Young in the early 19th century, where interference of light was evident and hence gave strong believe in the wave theory. In his famous work published in 1802 [7], he stated that:

> *"... wherever two portions of the same light arrive at the eye by different routes, either exactly or very nearly in the same direction, the light becomes most intense when the difference of the routes is any multiple of a certain length, and least intense in the intermediate state of the interfering portions; and this length is different for light of different colours."*

For many years afterwards, several physicians and mathematicians contributed to building solid foundations for the wave picture of light. Among them, a great credit goes to Fresnel (1788 – 1827), who could explain diffraction phenomena based on Huygens' theory. Calculations for the existence of light in the shadow region caused by diffraction at edges became thus possible. However, explanations for some other phenomena were not yet sufficiently available, e.g., polarization effects and light-matter interaction.

Parallel to the research on light, thorough investigations on electricity and magnetism were in progress. The physical laws governing electric and magnetic forces were observed experimentally and formulated mathematically by many scientists, like Ampère (1775 – 1836), Gauss (1777 – 1855), Faraday (1791 – 1867), and Weber (1804 – 1891). These quantitative formulations were first completed by the work of James Clerk Maxwell (1831 – 1879) after the introduction of the displacement current. He gathered his theories in his book "A Treatise on Electricity and Magnetism" published in 1873 [8]. By introducing the displacement current term [9] to the previously available equations describing the actions made by electric and magnetic fields, Maxwell formulated the fundamental four differential equations of electromagnetism, namely Maxwell's equations. In addition, the polarization of light could be explained due to the transversal wave nature of the propagating electromagnetic fields. This formulation could describe for the first time that electromagnetic signals can propagate in space with a velocity exactly equal to the experimentally known speed of light. It was then evident that light itself, as the visible part of the electromagnetic spectrum, is also of the same wave nature.

A few years after the foundation of Maxwell's equations, Heinrich Hertz (1857 – 1894) examined the propagation of electromagnetic waves and confirmed experimentally its speed to the one predicted by Maxwell and measured for optical signals. In his paper [10], the third conclusion stated this finding clearly, as said in the following original German text:

> *"Die schon durch viele Wahrscheinlichkeitsgründe gestützte Hypothese, dass die Transversalwellen des Lichtes electrodynamische Wellen seien, gewinnt feste Grundlage durch den Nachweis, dass es wirklich electrodynamische Transversalwellen im Luftraume gibt, und dass diese sich mit einer der Geschwindigkeit des Lichtes verwandten Geschwindigkeit ausbreiten. Auch öffnet sich ein Weg, jene wichtige Anschauung endgültig zu bestätigen oder zu widerlegen. Denn es scheint die Möglichkeit gegeben, die Eigenschaften electrodynamischer Transversalwellen auf dem Wege des Versuchs zu studiren und dieselben mit den Eigenschaften der Lichtwellen zu vergleichen."*

It translates into English [11] as:

> *"There are already many reasons for believing that the transversal waves of light are electromagnetic waves; a firm foundation for this hypothesis is furnished by showing the actual existence in free space of electromagnetic transversal waves which are propagated with a velocity akin to that of light. And a method presents itself by which this important view may finally be confirmed or disproved. For it now appears to be possible to study experimentally the properties of electromagnetic transversal waves, and to compare these with the properties of light waves."*

Following the advances in physics during the early 20th century, light was declared to have a duality of wave and particle nature. The experiments of Max Plank (1858 – 1947) on the black body radiation led to the postulation of quantized energy particles, which were later named photons. Each of these particles has a fixed energy amount depending on the frequency of radiation. This frame of new theories was widely developed in the 20th century, and became commonly known as the modern physics. Working at frequencies much lower than the visible spectrum, and at relatively high energy levels of radiation, the duality nature can be simplified to the wave nature well described by Maxwell's equations.

It cannot be said that the actual nature of light is fully discovered. Light, as an electromagnetic radiation, is still an open field for research, where science and philosophy sometimes meet. The frame of modern physics keeps expanding and searching for answers to questions regarding the interaction between light and matter. In spite of this, the description of light is well established for many applications with sufficient accuracy, which allows for explorations to new opportunities for the humankind by making use of the already discovered and quantified physical mechanisms and phenomena.

## 1.2 Imaging with electromagnetic waves

Since the understanding of light as an electromagnetic wave and the consequent recognition of the basic rules governing its propagation properties, many efforts have been conducted to utilize the electromagnetic theories for imaging purposes. Imaging is understood as a virtual reproduction of an actual object. The first steps were targeting the optical domain, namely for the visible spectrum. A remarkable milestone for the success of the electromagnetic theories in inventing new imaging methods is credited to Dennis Gabor (1900 – 1979), leading to the Nobel Prize in Physics in 1971. He introduced a concept for recording the wavefront reflected by an imaged object in a form named a hologram, which can be used afterwards to produce images from various looking angles [12]. The concept of hologram relies mainly on the spatial coherence of the light wavefront.

Spatial coherence helped a lot in the foundations of the electromagnetic wave theory of light, namely by the observation of interference and diffraction effects. Although spatial coherence can relatively easily be achieved in the optical domain, temporal coherence is much more complicated. It was not until the invention of the laser in the 1960's for the optical sources to achieve temporal coherence. In the lower part of the spectrum, i.e., at the microwave frequencies, however, the wavelengths are much longer than in the optical domain, which eased the realization of temporally coherent sources in practice. Encouraged by this technical advantage, full coherent imaging methods in the microwave range had been extensively developed during the 20th century. Combining spatial coherence and temporal coherence leads to a complete knowledge about the signal wavefronts, hence results in a full description for the association between the imaged object and the scattered waves. Accordingly, imaging quality can be enhanced significantly. In the upper part of the spectrum, i.e., at X-rays, the realization of temporally coherent sources is still suffering from a high level of technical difficulties. Alternatively, imaging concepts had been developed based on incoherent operation, e.g., X-ray computed tomography. Research, however, on temporally coherent X-ray sources [13], also named X-ray laser, is currently advancing.

The progress in microwave imaging is strongly influenced by the advances in radar systems. Radar is the elementary front-end part of the imaging system, and hence any technological enhancements in radar performance or miniaturization could directly assist the imaging capabilities. Range radar for navigation and remote sensing applications was enormously developed during the last decades [14], and is still in continuous improvement by pushing to higher frequencies with higher bandwidths, e.g., to the THz range [15]. Most of the radar systems operate in monostatic configuration, in which the transmitter and the receiver are co-located, or slightly separated for better isolation. This configuration is often chosen due to practical constraints, e.g., mechanical, connectivity, or cost issues. Advanced methods based on bistatic radar did not appear at first, however, later during the second world war. It is credited to the Germans, who made use of the transmitters of the British air defense system placed along the east cost of England as illumination sources to their receivers. The German system was called "Klein Heidelberg", and is well documented in [16] and the references therein. Intelligently, the Germans installed passive receivers on the opposite shore, and synchronized them to the pulsed transmission of the British antennas. With which, it was feasible to detect the distance to targets through the bistatic reflections, and hence to construct an ambiguity ellipse curve. Combining this with mechanical scanning of the receiver antenna in order to find the angle of arrival using a maximum peak technique, the target distance and location could be simultaneously identified. Although the system was principally capable to operate in multistatic configuration due to the availability of several transmitters and receivers, it would have been a major difficulty to establish a fast communication between all receivers. Nevertheless, this development sat a start for the bistatic radar applications.

Imaging with microwave frequencies, often denoted as the range from 0.3 to 300 GHz or alternatively wavelengths between 1 m and 1 mm, offers penetration capabilities for non-metallic materials. As the corresponding wavelengths are relatively long, achieving a good resolution is only possible when large apertures are also used. The construction of such mechanically large apertures had hindered the development of microwave imaging systems. For instance, when imaging with a signal of 1 GHz, the wavelength in free space is almost 30 cm. In order to achieve a resolution of 15 cm on 1 km distance, an aperture of around 2 km is required. This is beyond the mechanical producibility in practice. To overcome this constraint and encouraged by the astrophysical applications, the first concepts for synthetic aperture were developed in the 1950's for passive observation of the sky. This hence received high recognition in the scientific community and led to the Nobel Prize in Physics in 1974 to Martin Ryle and Antony Hewish. In the Nobel lecture of Ryle [17], one reads:

> *"Suppose now that only a small part of the wavefront is sampled, but that different parts are sampled in turn ...; could we combine these signals to produce the same effect? Since in general, we do not know the phase of the incident field at different times this would not normally be possible but if we continue to measure one of the samples while we measure the others we can use the signal from this one as a phase reference to correct the values measured in other parts of the wavefront.*

> *In this way, by using two small aerial elements, we can again add the fields over the wavefront – the area of which is now determined by the range of relative positions taken by the two aerial elements."*

The usage of a reference receiver was a necessity due to the lack of the temporal coherence in the received signals from the sky. On the contrary, spatial coherence is available and is used to spatially correlate the received signals over a large area, and then recording them, and afterwards coherently process them to synthesize a large aperture. This technique is thus named a synthetic aperture radar (SAR). The realization of this imaging technique is highly dependent on the availability of suitable recording medium and sufficient computation power. The SAR processing was initially achieved in the 1960's using optical recording mediums and optical processors. With the tremendous development in digital computing, this was later fully replaced by modern electronic recording mediums, where higher speeds and signal qualities are easier available. Nowadays, SAR is widely utilized in airborne and spaceborne imaging systems [18], mainly in monostatic configuration. Recently, bistatic acquisition became feasible after the advances achieved in the signal synchronization between satellites and their position tracking with the Global Positioning System (GPS), e.g., TanDEM-X satellite imaging system [19–21].

Electromagnetic wave theory proved to offer a theoretical frame for the development of microwave imaging systems. Microwave imagers are classified to operate in far-field or in close range, depending on the aperture size relative to the imaged distance. They can also be active or passive imagers depending on the source of illumination, whereas active ones have the capability for full coherent imaging, i.e., spatial and temporal.

## 1.3 Multistatic imaging

Among the various known classes of imaging systems, the most rich one is surely the multistatic imaging operating at close ranges. Multistatic leads to significant diversity in the correlated information about the scattering of the imaged object. Additionally, it indicates a high degree of flexibility in the choice of the transmitting and receiving apertures. This increased flexibility is thus combined with consequent complexity of the design process, in which many challenging decisions about the imaging system architecture and its array configuration are to be taken. In addition, a close range operation enhances the visibility of the object and allows for imaging with high signal-to-noise ratio.

Multistatic operation indicates active imaging by definition. Transmitters are used to illuminate the imaged object from different directions, and receivers collect the scattered field from other directions. For many imaging problems, active illumination is a must in order to ensure high dynamic range and independence from the surrounding illumination. When image processing techniques are demanded, the image dynamic range becomes an important success factor for stable operation. Microwave imaging is mainly used for its penetration capability of many optically opaque surfaces. However visibility through many of these objects is possible, they, nevertheless, cause signal attenuation. Active illumination is hence essential to let the signal reach behind the surface of the object with some decent power level. Utilizing active illumination also means that some issues can arise regarding specular reflections and shadowing effects. Understanding of the nature of these phenomena is thus important to either avoid them or reduce them depending on the application in question. The flexibility offered in the synthesis of multistatic arrays allows for a dedicated geometrical adaptation to match the imaged object, and thus to optimize the imaging performance.

The geometrical diversity delivered by multiple viewing of each single voxel of the imaged object enables access to a large portion of the spatial frequency domain. This thus results in a better utilization of the electromagnetic spectrum, i.e., better use of natural resources. The multiple views translate to a large spreading in the spatial spectrum, which in turn leads to an enhancement in image resolution and artifacts rejection. While combining the multiple views with a wide frequency band, an extended allocation in the spatial spectrum is achieved, thus explains the great features of multistatic imaging. This puts multistatic imaging in the class of optical imaging with lenses, for which each portion of the lens is naturally permitted to emit and to receive light.

Image reconstruction is the heart of any synthetic focusing system. In order to gain the best of performance from multistatic arrays, synthetic focusing is essential. Adapting to modern technologies, signal digitization and digital processing of data is a paradigm shift in radar imaging systems. Hence, digital-beamforming technique is a keystone in the success of this modern technology. However the aperture is not necessarily synthetic, as in airborne and spaceborne SAR systems [22], signal focusing is preferably kept based on synthetic aperture approach in order to benefit the high flexibility and at the same time to significantly reduce the hardware complexity of the RF parts. This reflects also a reduction of cost and a possibility to access higher frequency bands, where complex circuitry is often hard to build due to space and thermal constraints. Thanks to the availability of high computation power nowadays, realization of relatively complex and lengthy numerical computations became feasible and affordable. This offers many chances for multistatic imaging systems, which thought to be impractical and even impossible some years ago.

Fourier imaging methods had been established since many years to assist microwave imaging techniques where synthetic focusing is realized. Being demanded in the applications of airborne and spaceborne SAR systems [23], mathematical developments covering monostatic configurations had been presented and some are still in progress. Imaging with multistatic arrays can benefit from this knowledge in that a similar systematic description of imaging properties is feasible. In addition, comparison to other known techniques, e.g., monostatic imaging, can be conducted. In analyzing the performance of multistatic systems, however, both signal representations have to be addressed, namely in the space domain and the spatial frequency domain. Some phenomena are better explained by studying one domain rather than the another, hence considering both builds a solid understanding about multistatic imaging operation.

Though multistatic arrays can accommodate to any geometry, planar multistatic arrays are dealt with specifically because of many practical aspects. Planar arrays thus means that reflection imaging is considered, i.e., illumination and reception are both on the same side relative to the imaged object. For some applications, transmission imaging could be more favorable than reflection imaging. Many practical scenarios, however, include objects which are not fully transparent for microwave signals, hence would cause strong shadows in transmission imaging. Reflection imaging, instead, offers penetration through the outer layers of the object and can still receive the signals from the same side. Furthermore, planar systems are much more compact compared to the systems imaging in transmission mode. Therefore, the developments introduced here are solely targeting the case of planar multistatic arrays; it is, however, possible to extend the discussion in many parts to a general conformal array geometry.

With planar multistatic arrays, three-dimensional imaging is achieved. The two-dimensional lateral focusing is obtained due to the lateral extensions of the array itself; hence focusing is mainly gained by conventional spatial Doppler phenomenon [24]. The third dimension, namely the range direction, is obtained due to two phenomena, pulse compression and angular diversity. Pulse compression is a common technique in radar processing [25], which is utilized to achieve time or range gating of the signals, e.g., using matched filtering method. In addition to this, multistatic arrays at close ranges offer an exceptional high level of angular diversity and hence also a relatively deep allocation of the spatial spectrum. Both phenomena lead to an enhanced range focusing. This imaging method hence yields fully focused three-dimensional complex-valued images. Phase information is thus maintained, which is very useful in extracting features about the imaged object depending on the absolute frequency of operation rather than the signal bandwidth utilized. This accordingly encourages imaging at high frequencies, e.g., millimeter- and sub-millimeter-waves.

## 1.4 Potential applications

The capabilities of multistatic imaging can serve in many applications. They can also be an efficient replacement for other technologies currently utilized in microwave imaging systems, e.g., monostatic imagers, scanning phased arrays, or mechanical scanners. However new applications are still expected to evolve, recent developments demonstrate the applicability of microwave imaging in several domains. For instance, various imaging systems have been already developed to address the application of personnel screening with millimeter-waves, especially for airport security check points. Being a non-ionizing radiation, it is favorable to deploy millimeter-wave imagers instead of X-ray methods [26] in screening passengers. They are also superior to the conventional solutions based on metal detector gates, because of their capability to detect dielectric objects [27, 28], e.g., plastic explosives and ceramic knifes, besides other metal objects. Some devices are already commercialized and many still in development. They range from passive to active, and also from close range to stand-off imagers [29–46]. Fig. 1.1 illustrates an example for the usage of multistatic imaging in detection of a concealed pistol. This example is taken from the results introduced in the following chapters.

The powerfulness of microwave imaging extends also to applications considering material characterization and non-destructive testing (NDT). The interest is rapidly increasing to establish microwave imaging methods, where contact-less examination of bulk objects or raw materials is possible. For this, dedicated frequency ranges have to be identified according to the application in question. General purpose imaging solutions are also considered in order to assist research issues. It is, therefore, important when analyzing theories on microwave imaging to keep enough level of generalization to help the acquired knowledge to be adaptable in different situations. During the last years, many development results were already reported [47–53], and even some more are pushing towards the THz domain [54–56]. Fig. 1.2 represents a realistic example for the gained visibility through a bulky object in the millimeter-wave range.

It is to emphasize that this new technology is considered in its first steps. Research is ongoing to advance the imaging method on one hand, and on the other hand to try to spot potential applications in need of the imaging performance of multistatic arrays. For which, it is essential to deepen the understanding about its points of novelty. Planar multistatic arrays should better address applications where high resolution imaging with high image dynamic range is demanded. They are more immune against aliasing effects and are capable to operate indoors as well as outdoors without a need for an additional illumination source, e.g., the sun. Last but not least, three-dimensional complex-valued imaging is offered.

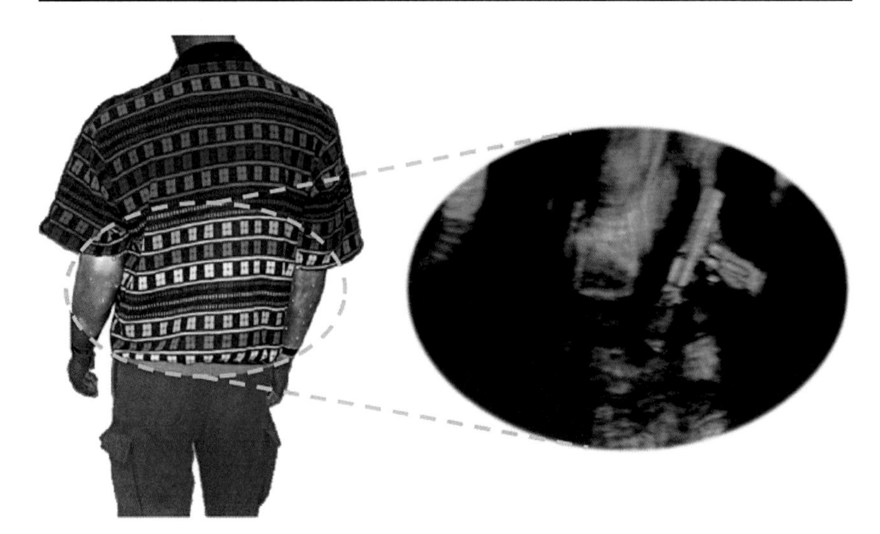

Figure 1.1: An example for microwave imaging in personnel screening applications. The microwave image on the right is achieved using a frequency range of 70 to 80 GHz from a distance of one meter, with which the concealed pistol is clearly revealed.

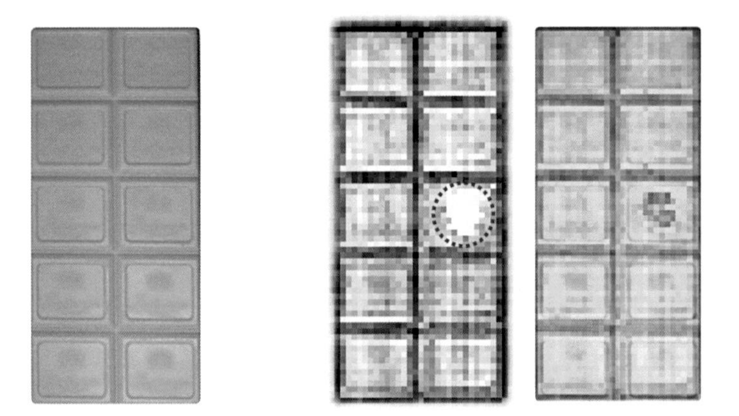

Figure 1.2: An example for microwave imaging in non-destructive testing applications. A white chocolate bar of 18 cm times 8 cm in size and around 1 cm in thickness has a hidden metal piece in the middle right section. An optical image is seen on the left, whereas on the right the microwave image is presented along with an overlay representation of the optical and the microwave images. The image is made in reflection arrangement using the same frequency range as above. The metal piece is successfully discovered, and is indicated in red color.

## 1.5 Challenges

The realization of the above mentioned features to advance microwave imaging quality based on planar multistatic arrays is exceptionally challenging. It requires a strong merging of digital solutions seamlessly with the analog front-ends. The design requirements of the analog front-ends must offer a possibility to overcome the conventional cost barrier faced when realizing arrays assembling hundreds or thousands of RF channels. Some modern semiconductor technologies offer good performance with yet moderate cost, thus opening the chance for realizing large multistatic arrays. Meanwhile, digital solutions are getting remarkably faster in speed and by parallelization, thus allowing the integration of many floating-point processors in a still manageable and affordable integrated solution. In order to utilize these modern and future technologies for imaging purposes, theoretical foundations for new imaging methods are thus required, which are to be conceived to efficiently benefit from this level of integrity.

Putting all these challenges together, the following lists some main target points in order to realize a microwave multistatic imager operating at close range and capable to address various applications, e.g., personnel screening and non-destructive testing.

- Lateral resolution:

  For various applications, it is demanded to reach lateral resolutions of a few millimeters on a distance of approximately one meter. This ensures enough visual details sufficient for analyzing or detecting many objects or to identify defects. Such a requirement has a direct influence on the aperture size of the imaging array and its frequency band of operation. Imaging with large apertures at close ranges means a consequent challenge in the design of the antennas, which have to offer very wide opening angles, however with stable phase centers and good isolation. The frequency band is favorably chosen at the upper band of the millimeter-wave range, which enhances the resolution while a significant level of penetration is still available.

- Range resolution:

  Resolving objects along range direction is highly beneficial in analyzing layered objects. The valuable three-dimensional imaging can only be maintained when range resolution comparable to the lateral one is available. Being to a large extent a product of the signal bandwidth used, usage of high frequency bands is encouraged where signal bandwidths are technologically easier to guarantee. Imaging operation with a signal bandwidth of several GHz is striven in order to reach range resolution of some millimeters.

- Image dynamic range:

  Image interpretation gets disturbed by noise and artifacts. Image dynamic range is one of the critical factors defining the imaging quality. It is limited by hardware related issues as well as the theoretical limit given by the array itself. Hardware limits image quality according to available output power, noise figure of the receiver, signal stability of the RF sources, quantization effects, and numerical rounding errors. Some aspects can be traded-off by increasing costs, however others are limited by the state-of-the-art technologies. Image dynamic ranges exceeding 30 dB are set as a favorable level of quality, with which many advanced image processing algorithms for image filtering or object classification can run robustly.

- Illumination quality:

  Similar to the optical domain, active imaging can experience problems due to strong specular reflections. It reacts like photographing of mirroring surfaces with a strong flash, thus producing very bright spots and strong shadows. This is critical to some applications, where object surfaces are electrically smooth. In order to overcome this difficulty, the array design considers homogenous and equal illumination, which thus requires further processing steps to correct signal levels.

- Technology:

  In order to first examine and later to demonstrate all introduced methods, it is essential to decide for a technological realization. For this, the millimeter-wave frequencies are chosen. The large number of RF channels demanded by this imaging method is better served by a well-established semiconductor technology. Off-the-shelf products are not suitable in this case, and, instead, effort is put into a customization of a silicon-germanium (SiGe) chip-set originating from the automotive industry in order to specifically support the imaging applications. For this, frequency operation around 75 GHz is used throughout all analysis and evaluations of the introduced imaging method. The results are, of course, not limited to this frequency range and can be easily adapted to other demands.

- System stability:

  Dealing with an RF system including thousands of channels, which are to operate synchronously in phase, is a great challenge. Accurate phase measurement is crucial for synthetic focusing, and hence for the whole system operation. Consequently, a lot of care must be taken in the hardware realization in order to guarantee stable and reproducible measurement acquisition. Additionally, efficient system calibration is an essential step to correct for systematic errors and to establish a reference phase plane for the imaging array.

- Power consumption:

  Despite the many advances offered by modern integration technologies, it is, nevertheless, important to account for power consumption and heat dissipation in the system. Many trivial array geometrical solutions are prohibited by thermal constraints, for instance. Antenna positioning within the array aperture should avoid dense placements, which might lead to poor RF isolation and excessive heat dissipation. On the contrary to mechanical scanning systems with virtual antenna positions, array antennas will be attached in practice by actual hardware components needing space for integration and sufficient thermal connectivity for heat dissipation.

- Speed:

  Fully electronic imaging is decided. It is a necessary step to advance the quality of operation made by microwave imagers. Mechanical scanning is fully avoided due to the associated tolerances and uncertainties. Although fully electronic measurement acquisition is applied, image blurring can occur once a movement of a fraction of the wavelength during the measurement time happens. Hence, effort is also put into fast measurement times. In practice, image reconstruction time could be also critical. It thus considerably influences the frame rate offered by the imaging system. However still being a bottleneck in system performance, current advances in digital computers and integrated digital processing units guarantee significant potential for continuous improvement in this regard.

- Cost:

  Being an engineering problem, aspects for cost reduction are considered where possible. The main key factor for this is definitely the total number of antennas used in an imager. Multistatic arrays prove to be superior in this context due to the huge reduction factors they can offer. Their geometry is to be selected as sparse as possible. This leads to a typical trade-off situation between quality and cost, which is resolved by the introduced techniques in order to reach a best compromise.

## 1.6 Methodology

Solid understanding of the operation of multistatic arrays from the very fundamental relations of electromagnetic waves is first developed. Common terminologies from other imaging domains, e.g., optics, Fourier imaging, seismic imaging, synthetic aperture radar, and ultrasonic, are adapted where possible, and extended when required. Purposely, heavy use of digital signal processing is utilized in order to reach the best possible performance from the analog front-ends. Similarly, focusing of the signals is conceived based on accurate phase terms, thus avoiding any typical simplifications known historically in radar development. This assumes a strong degree of unification between the analog and digital parts of the imaging system, which hence meets the realistic situation of future imaging devices.

Experimental or numerical validations follow every mathematical development in order to examine its correctness based on realistic evaluation scenarios. Additionally, practical aspects are greatly discussed, in which the influences of the considered assumptions and the associations of the applied design choices become clear. Care is everywhere taken in preparing theoretical solutions feasible to be applied and realized in real systems. This is done, however, while keeping a significant degree of generalization in order to avoid application specific solutions.

The foundations of the planar multistatic arrays are first detailed in chapter 2, with the illustrations always assigning the red color for transmitters and the blue one for receivers. This color code is kept for convenience throughout all following chapters. The problem of object illumination in active imaging is discussed comprehensively in chapter 3, followed by a specific analysis of the illumination on the human body. Chapter 4 addresses the calibration problem of planar multistatic arrays, which is a necessity for a proper operation of microwave imaging systems. Following the achieved results, a fully electronic demonstrator based on an introduced novel planar multistatic imaging array is presented in chapter 5, with which imaging of humans is also experimented. Additionally, various imaging results are reported in the appendix image gallery, through which a deeper understanding of the imaging capabilities of the millimeter-waves is considerably attained.

Chapter 2

# Multistatic Array Imaging

---

Imaging with planar multistatic arrays is studied in this chapter starting from the fundamental relations of electromagnetics to the practical aspects associated with the realization of complex large arrays. Purposely, the study is generalized in order to include all possible array geometries.

The development of the electromagnetic field relations follows to some extent the approach made by Born and Wolf in [57]. While Born's work was more dedicated to the optical domain, it is necessary to adapt this to the notations used in microwave theories. Furthermore, the analysis extends to include the method of Fourier array imaging in order to describe the image properties made by planar arrays. The notations used by Soumekh in [23] for the analysis of one-dimensional arrays are extended to cover multistatic array imaging with two-dimensional arrays. Accordingly, three-dimensional imaging is achieved, therefore the emphasis on the three-dimensionality is taken into account throughout all mathematical formulations. Image reconstruction methods based on space domain representation as well as spatial frequency domain one are also formulated. Moreover, the development is combined with discussions on the practical aspects affecting the image properties in terms of spatial resolutions, point spread function, and illumination coverage.

Sparse array design is essential to enable the realization of fully electronic imaging arrays. While synthesis of multistatic arrays operating at close ranges is very challenging, a suitable method based on effective aperture technique is discussed in order to assist the synthesis problem to produce arrays of high sparseness, nevertheless delivering high image quality. Design examples combined with simulation results as well as experimental verifications are presented to prove the applicability of the synthesis method and to examine the imaging performance of the arrays. Lastly, a design of a large planar multistatic array with a one square meter aperture is addressed in detail and is used to verify quantitatively the theoretical results presented earlier. The array is then used to evaluate a realistic test scenario for personnel screening application.

## 2.1 Electromagnetic field relations

The fundamental relations of electromagnetism are given by the four Maxwell's equations. The equations in their integral representation are expressed as

$$\oint_C \mathscr{E} \cdot d\ell = -\frac{\partial}{\partial t} \iint_S \mathscr{B} \cdot \mathbf{ds} \tag{2.1}$$

$$\oint_C \mathscr{H} \cdot d\ell = +\frac{\partial}{\partial t} \iint_S \mathscr{D} \cdot \mathbf{ds} + \iint_S \mathscr{J} \cdot \mathbf{ds} \tag{2.2}$$

$$\oiint_S \mathscr{D} \cdot \mathbf{ds} = \mathscr{Q} \tag{2.3}$$

$$\oiint_S \mathscr{B} \cdot \mathbf{ds} = 0 \tag{2.4}$$

where $\mathscr{E}$, $\mathscr{D}$, $\mathscr{H}$, $\mathscr{B}$, and $\mathscr{J}$ denote the three-dimensional vector quantities of the electric field, the electric flux density, the magnetic field, the magnetic flux density, and the electric current density, respectively; and $\mathscr{Q}$ denotes the electric charge as a scalar quantity. The symbols $C$ and $S$ emphasis linear and surface integrals, respectively. The circles around the integrals indicate closed integrations either over a contour or a surface. For contour integrals, the chosen directions of the path $C$ and the corresponding unit vector $d\mathbf{s}$ must follow the right-hand rule; and for a closed surface integral, $d\mathbf{s}$ must point outwards. $\mathscr{E}$, $\mathscr{D}$, $\mathscr{H}$, $\mathscr{B}$, $\mathscr{J}$, and $\mathscr{Q}$ exhibit arbitrary dependencies on space and time $t$; and are said to be a solution for an electromagnetic problem if and only if they satisfy Maxwell's equations.

Due to the physical fact that the charge is conserved in nature, a supplementary equation relates the electric current density to the electric charge itself by

$$\oiint_S \mathscr{J} \cdot \mathbf{ds} = -\frac{\partial}{\partial t} \mathscr{Q} \,, \tag{2.5}$$

which is also named the continuity equation.

However the integral representation of Maxwell's equations is more intuitive to understand, it is very helpful to rewrite them in a differential representation. This alternatively describes the behavior of the field components locally, thus expressed as

$$\nabla \times \mathscr{E} = -\frac{\partial}{\partial t}\mathscr{B} \tag{2.6}$$

$$\nabla \times \mathscr{H} = +\frac{\partial}{\partial t}\mathscr{D} + \mathscr{J} \tag{2.7}$$

$$\nabla \cdot \mathscr{D} = \rho \tag{2.8}$$

$$\nabla \cdot \mathscr{B} = 0 \tag{2.9}$$

with the continuity equation also rewritten as

$$\nabla \cdot \mathscr{J} = -\frac{\partial}{\partial t}\rho \,, \tag{2.10}$$

where $\rho$ denotes the electric charge density.

Maxwell's equations are a linear set of differential equations and hence the solution of time-harmonic signals can be expressed by a phasor representation in order to eliminate the time dependency and simplify the mathematical formulations. The relation between any vector field $\mathscr{A}$, with a time-harmonic variation of an angular frequency $\omega$, and its corresponding phasor $\mathbf{A}$ is

$$\mathscr{A}(\mathbf{r}, t) = \Re\left\{\mathbf{A}(\mathbf{r})\, e^{j\omega t}\right\} \,, \tag{2.11}$$

where $\Re\{\,\}$ denotes the real part. Accordingly, any time derivative on $\mathscr{A}$ corresponds to a multiplication factor of $j\omega$ in the phasor representation.

Maxwell's equations in the differential representation can now be rewritten using phasors as

$$\nabla \times \mathbf{E} = -j\omega\mathbf{B} \tag{2.12}$$

$$\nabla \times \mathbf{H} = +j\omega\mathbf{D} + \mathbf{J} \tag{2.13}$$

$$\nabla \cdot \mathbf{D} = \rho \tag{2.14}$$

$$\nabla \cdot \mathbf{B} = 0 \tag{2.15}$$

with the continuity equation similarly rewritten as

$$\nabla \cdot \mathbf{J} = -j\omega\rho \,. \tag{2.16}$$

This formulation is also named as the frequency domain representation of Maxwell's equations.

For an isotropic linear inhomogeneous material with permittivity $\varepsilon(\mathbf{r})$, permeability $\mu(\mathbf{r})$, and conductivity $\sigma(\mathbf{r})$, the constitutive equations relating the material electrical characteristics are defined by

$$\mathbf{D} = \varepsilon(\mathbf{r}) \cdot \mathbf{E} \tag{2.17}$$

$$\mathbf{B} = \mu(\mathbf{r}) \cdot \mathbf{H} \tag{2.18}$$

The value of $\varepsilon(\mathbf{r})$ is generally complex, which includes the material losses due to its conductivity; hence can be expressed as

$$\varepsilon(\mathbf{r}) = \varepsilon_0\varepsilon_r(\mathbf{r}) + \sigma(\mathbf{r})/j\omega \tag{2.19a}$$

$$= \varepsilon_0\varepsilon_r(\mathbf{r})\left(1 + \frac{\sigma(\mathbf{r})}{j\omega\varepsilon_0\varepsilon_r(\mathbf{r})}\right) \tag{2.19b}$$

$$= \varepsilon_0\varepsilon_r^c(\mathbf{r}) \tag{2.19c}$$

with $\varepsilon_0$, $\varepsilon_r$, and $\varepsilon_r^c$ denote the vacuum permittivity, the real part of the relative permittivity, and the complex relative permittivity of the material, respectively. The material characteristics represented by $\varepsilon_r(\mathbf{r})$, $\mu(\mathbf{r})$, and $\sigma(\mathbf{r})$ are generally also frequency dependent.

## 2.2 Scattering in a source-free region

In the following, the scattering problem in a source-free region is analyzed, where the focus is made on the interaction between the incident field on the material, and the accordingly scattered field. A source-free region is given where $\mathbf{J} = 0$ and $\boldsymbol{\rho} = 0$. Hence by using the constitutive equations, the field relations are reduced to

$$\nabla \times \mathbf{E} = -j\omega\mu(\mathbf{r})\mathbf{H} \tag{2.20}$$

$$\nabla \times \mathbf{H} = +j\omega\varepsilon(\mathbf{r})\mathbf{E} \tag{2.21}$$

$$\nabla \cdot (\varepsilon(\mathbf{r})\mathbf{E}) = 0 \tag{2.22}$$

$$\nabla \cdot (\mu(\mathbf{r})\mathbf{H}) = 0 \tag{2.23}$$

It is to be noticed that scattering from highly conductive materials, e.g., metals, cannot be dealt with under a source-free condition due to the arising surface currents and charges at the boundaries. Therefore, the analysis is restricted to dielectric materials of possibly low conductivity. Deviation from this assumption, and the resultant consequences for an imaging system, will be discussed later in chapter 3.

In the absence of sources, eq. 2.22 and 2.23 indicate a solenoidal behavior of the modulated $\mathbf{E}$ and $\mathbf{H}$ fields by the spacial variations of the permittivity and the permeability of the immersed material. The equations 2.20 through 2.23 represent coupled differential equations in the vector fields $\mathbf{E}$ and $\mathbf{H}$. Taking into account the spatially varying $\varepsilon$ and $\mu$, the differential equations can be separated as shown next.

The calculations make use of the following general vector identities for a scalar field $\alpha$ and a vector field $\mathbf{\Psi}$

$$\begin{aligned}
\nabla \times (\alpha \mathbf{\Psi}) &= \alpha(\nabla \times \mathbf{\Psi}) + (\nabla \alpha) \times \mathbf{\Psi} && (2.24)\\
\nabla \times \nabla \times \mathbf{\Psi} &= \nabla(\nabla \cdot \mathbf{\Psi}) - \nabla^2 \mathbf{\Psi} && (2.25)\\
\nabla \cdot (\alpha \mathbf{\Psi}) &= (\nabla \alpha) \cdot \mathbf{\Psi} + \alpha(\nabla \cdot \mathbf{\Psi}) && (2.26)\\
(\nabla \alpha)/\alpha &= +\nabla(\ln(\alpha)) && (2.27)\\
\alpha \nabla(1/\alpha) &= -\nabla(\ln(\alpha)) && (2.28)
\end{aligned}$$

By dividing eq. 2.20 by $\mu(\mathbf{r})$, and then applying a curl operation to both sides, it becomes

$$\nabla \times \left( \frac{1}{\mu(\mathbf{r})} (\nabla \times \mathbf{E}) \right) = -j\omega \, \nabla \times \mathbf{H} \, . \qquad (2.29)$$

Eq. 2.21 is then substituted in eq. 2.29 in order to replace the $\nabla \times \mathbf{H}$ term; and after applying the identity of eq. 2.24 it yields

$$\nabla \times \nabla \times \mathbf{E} + \Big( \mu(\mathbf{r}) \nabla \Big( 1/\mu(\mathbf{r}) \Big) \Big) \times (\nabla \times \mathbf{E}) = \omega^2 \varepsilon(\mathbf{r}) \mu(\mathbf{r}) \mathbf{E}. \qquad (2.30)$$

The application of eq. 2.22 through the identities of eq. 2.26 and eq. 2.27 yields

$$\nabla \cdot \mathbf{E} = (\nabla \ln \varepsilon(\mathbf{r})) \cdot \mathbf{E}. \qquad (2.31)$$

The identity of eq. 2.25 is now applied to the $\nabla \times \nabla \times \mathbf{E}$ term in eq. 2.30, and then eq. 2.31 is substituted in the result, hence after some mathematical rearrangements, it yields

$$\nabla^2 \mathbf{E} + \omega^2 \varepsilon(\mathbf{r}) \mu(\mathbf{r}) \mathbf{E} = \Big( \nabla \ln \mu(\mathbf{r}) \Big) \times (\nabla \times \mathbf{E}) - \nabla \Big( (\nabla \ln \varepsilon(\mathbf{r})) \cdot \mathbf{E} \Big) \, . \qquad (2.32)$$

Eq. 2.32 describes the time-harmonic electric field in any source-free isotropic linear inhomogeneous medium. The right hand side of the equation couples the three field components in Cartesian coordinates, and hence complicates the solution. Further restrictions on the type of material are made next in order to allow for a simplified expression of the scattering process.

- In practice, a wide range of dielectric materials does not show significant magnetization effects, for which $\mu(\mathbf{r})$ can be assumed constant and equal to $\mu_0$, i.e., the vacuum permeability. Then the first term on the right hand side vanishes.

- The spatial derivatives of $\varepsilon(\mathbf{r})$, and similarly for the $(\ln \varepsilon(\mathbf{r}))$, are relatively low compared to the field $\mathbf{E}$. Hence, as a matter of approximation, the second term of the right hand side can be neglected in comparison to the left hand side.

- The material will be considered non-dispersive within the frequency range of observation. In microwave and millimeter-wave ranges, the observation bandwidths are of some GHz, within which the permittivity of the material can be considered constant without significant error.

Thus, eq. 2.32 can be reduced to the homogeneous Helmholtz's equation

$$\nabla^2 \mathbf{E} + k_0^2 \varepsilon_r^c(\mathbf{r})\mathbf{E} = 0\,, \tag{2.33}$$

where $k_0 = \omega\sqrt{\varepsilon_0 \mu_0}$ is the wavenumber in free space. Alternative expressions for $k_0$ are

$$k_0 = \frac{\omega}{c_0} = \frac{2\pi f}{c_0} = \frac{2\pi}{\lambda_0}\,, \tag{2.34}$$

where $c_0$, $f$, and $\lambda_0$ denote the speed of light in free space, the signal frequency, and the signal wavelength in free space, respectively.

The Cartesian field components of $\mathbf{E}$ in eq. 2.33 are no more coupled, and hence the equation can be split into three independent ones for each field polarization. This suggests the replacement of the vector field $\mathbf{E}(\mathbf{r})$ with a scalar field $U(\mathbf{r})$ representing any of the three polarizations. Hence, eq. 2.33 is rewritten as

$$\nabla^2 U + k_0^2 \varepsilon_r^c(\mathbf{r}) U = 0. \tag{2.35}$$

The scalar field $U$ is composed of the incident $U^i$ and scattered $U^s$ fields, namely

$$U = U^i + U^s\,. \tag{2.36}$$

$U^i$ represents the field solution in the absence of scatterers, i.e., $\varepsilon_r^c(\mathbf{r}) = 1$, where from eq. 2.35 it follows

$$\nabla^2 U^i + k_0^2 U^i = 0\,. \tag{2.37}$$

Substituting eq. 2.36 and 2.37 into eq. 2.35, and after rearranging the result, it states that

$$\nabla^2 U^s + k_0^2 U^s = -O(\mathbf{r}) \cdot U\,, \tag{2.38}$$

where $O(\mathbf{r})$ is defined as

$$O(\mathbf{r}) = k_0^2(\varepsilon_r^c(\mathbf{r}) - 1)\,. \tag{2.39}$$

The function $O(\mathbf{r})$ was named the scattering potential by Born and Wolf [57, 58]. It is seen as the object function describing the variations of permittivity within the body of the scatterer, and thus contains its image.

The following formulations will target a solution for $O(\mathbf{r})$ within the frame of approximations discussed above.

Eq. 2.38 takes the form of an inhomogeneous Helmholtz's equation for the scalar field $U^s$ and with the source given by $-O(\mathbf{r}) \cdot U$. As commonly known, a suitable approach is the usage of the scalar Green's function to solve the Helmholtz's equation, thereafter the solution is expressed as a convolution integral. Considering $G(\mathbf{r})$ to be the solution for the three-dimensional Dirac delta function $\delta(\mathbf{r})$, namely

$$\nabla^2 G(\mathbf{r}) + k_0^2 G(\mathbf{r}) = -\delta(\mathbf{r}) \,, \tag{2.40}$$

it can be analytically shown, as in Appendix A, that the solution for the Green's function is given by

$$G(\mathbf{r}) = \frac{e^{-jk_0 r}}{4\pi r} \,. \tag{2.41}$$

This Green's function describes an outward propagating spherical wave out of the origin, where the source $\delta(\mathbf{r})$ is placed.

Now the solution of eq. 2.38 can be expressed by the volume convolution integral

$$U^s(\mathbf{r}) = \iiint_V O(\mathbf{r}') \cdot U(\mathbf{r}') \cdot G(\mathbf{r} - \mathbf{r}') \, d\mathbf{r}' \tag{2.42}$$

$$= \big(O(\mathbf{r}) \cdot U(\mathbf{r})\big) * G(\mathbf{r}) \,. \tag{2.43}$$

Eq. 2.43 states that the total scattered field is due to the superposition of the scattered waves at each volume portion of the scatterers. The excitation is driven by the total field $U$, which is weighted by the $O$ value and then radiates outwards the scatterers. The relation between the incident field $U^i$ and the scattered field $U^s$ in this representation is nonlinear.

## 2.3 Born approximation

The nonlinearity in eq. 2.43 hinders a direct inversion of the scattering equation. Therefore, a further approximation is to be applied in order to linearize the equation and hence allow for easier calculations. This approximation addresses the total field $U$ inside the convolution integral. For weak scatterers, the scattered field $U^s$ is negligible relative to the incident one $U^i$, hence

$$U|_{\text{low } \varepsilon_r^c} \approx U^i. \tag{2.44}$$

This approximation is widely known and named as Born approximation, or first order Born approximation, after it was used in the field of quantum mechanics by Born [59]. Higher order approximations are possible by iteratively solving the convolution integral as suggested in [57], however in many applications the first order approximation is considered sufficient for the range of accuracy required. Following eq. 2.44, the integral in eq. 2.42 reduces to

$$U^s(\mathbf{r}) = \iiint_V O(\mathbf{r}') \cdot U^i(\mathbf{r}') \cdot G(\mathbf{r}-\mathbf{r}')\, d\mathbf{r}' \qquad (2.45)$$

$$= \left(O(\mathbf{r}) \cdot U^i(\mathbf{r})\right) * G(\mathbf{r})\,. \qquad (2.46)$$

## 2.4 Validity of Born approximation

Following the Born approximation, the immersed material is considered to be of low dielectric constant and weak conductivity, and hence the solution supposes full penetration within its volume. This is not always the case in practice, as imaging of materials generally ranging from dielectrics to metals is demanded. The range of validity for the derived solution above is, however, restricted to the assumptions made for the material electrical characteristics as well as the simplifications considered regarding the scattering process. In spite of this, the deviation from these restrictions does not falsify the solution completely, however in many cases causes raising of systematic artifacts in the later reconstructed image. In order to understand the origin and behavior of these effects, the various phenomena influencing the range of validity are discussed below.

- Multiple reflections:

  Each volume portion of the scatterer would scatter part of the incident signal reaching it to the surrounding scatterers, which in turn will repeatedly scatter the signal. Due to the linearization made in eq. 2.46, this effect is completely neglected. When imaging strong scatterers, like metals, multiple reflections arise within the scatterer volume and can produce false scatterer positions, i.e., bright spots of actually non-existing objects [60]. In close range multistatic imaging, these spots are usually blurred and appear out-of-focus, because of the reduced possibility of coherent reconstruction at their positions resulting from the strong diversity in the view angles.

- Attenuation:

  When the material is partially conductive, and hence lossy, the signal propagating through it is attenuated, and consequently all imaged objects in its background would appear of less reflectivity than they really are. In addition to this, the Green's function solution of eq. 2.41 is often simplified in practice by dropping the space attenuation factor, due to reasons related to the numerical implementation of the reconstruction process. This also leads to a false estimate of the signal attenuation in relation to the distance to the antenna. Therefore, the local magnitude of the image at some position is understood to be relative to its neighborhood, rather than as an absolute value.

- Signal delay:

  The signal travel time through the body of the material is reduced according to the value of its dielectric permittivity. Again, due to the ignoring of the scattered field inside the scatterers, the variations in the propagation time is no more considered, and is however fixed to a constant value. Consequently, the image of a homogeneous dielectric appears to occupy larger space rather than it physically and geometrically does. This is beneficial to some applications targeting to identify dielectric objects from another surrounding material.

- Wave interference:

  Interference of incident and scattered waves happening within the imaged object cannot be accurately analyzed in the frame of the first order Born approximation. If the geometry of the scatterer is of some regular structure causing systematic interferences or some resonance modes, then the illumination in the image will appear inhomogeneously distributed under the Born approximation. This leads to a loss of information at the positions where the total field is nulled, or weakened by the interferences. In practice, applying a significant signal bandwidth can average out such effects and hence reestablish the illumination homogeneousness.

- Speckles:

  By coherent illumination of a rough surface with a roughness comparable to the applied wavelength, the scattered fields sum up in some chaotic manner and disturb the incident field significantly. The total field thus applied to the object is no more well described by the incident field. The simplification of the scattering process made on eq. 2.32 as well as by the Born approximation are both far inaccurate in such a case. This raises an interference between the incident and reflected waves causing speckles in the reconstructed image afterward. Speckles are spatially fast varying deterministic noise. This is usually annoying for any imaging system [61], and is mainly reduced by applying a considerable bandwidth to the signal illuminating the object. Consequently, the chaotic variations in the reflected wave add destructively over frequency, causing the speckles to vanish gradually with increased signal bandwidth. Similarly, the speckles get reduced when multiple views are applied while imaging the rough surface. Multistatic imaging with applied signal bandwidth is hence capable to avoid speckles to a significant extend.

- Specular reflections:

  When imaging smooth highly reflective surfaces, the reflected wavefront is coherent and includes energy transfer in a limited angle range relative to the incident wave direction. This is the known phenomenon of specular reflections addressed often in optical systems. Metallic surfaces as well as objects with high permittivity are typical examples for specular reflective objects. This type of reflection is strictly not covered by the presented solution, either due to the assumption of source-free region or the assumed low value of the relative dielectric constant. The phase information assumed by the presented solution has limited validity within a small angle variation around the ray following the law of reflection, i.e., the angle of reflection is equal to the angle of incidence. Detailed description of this phenomenon and its consequences on imaging electrically smooth objects will be addressed with greater detail in chapter 3.

- Shadowing:

  Shadows occur when blockage of the incident or reflected signal along their propagation paths exists. The approximation takes into account that the incident wave, for instance, will be able to reach all positions within the imaged object body. Materials of high permittivity or of some conductivity will block the propagation of the incident wave and hence result in dark zones in the image. The same applies for the receiving path.

- Polarization dependency:

  Polarization dependent scattering is strongly produced by edges and wedges. This effect is not fully addressed by the frame of scalar theory. Instead, indications of polarization dependent scattering can be observed when the object brightness varies in accordance to its lateral orientation relative to the applied signal polarization. The scalar theory, however, can be extended to address incident and reflected waves of two orthogonal polarizations, hence four type of scattering combinations would be formulated. This extension of information could help in some applications to reach a better classification of the scatterers.

In spite of the various restrictions the first Born approximation causes, it is proved to be very efficient in many imaging applications. In the specific cases when objects to be imaged are strongly surrounded by inhomogeneous dielectrics, then Born approximation could fail and hence better approximations must be considered [62–64].

## 2.5 The imaging problem

Imaging an object in a reflection arrangement is understood as the process of correlating the information of the reflected scattered field out of the object from multiple angles of view in order to find a description of the object function. Much similar to the operation principle of a simple focusing lens viewing the imaged object from a certain angle range, the digital-beamforming technique correlates the spatially sampled reflected field from several view angles, and optionally also frequencies, to reconstruct an image about the object. The correlation depends in this case on the scattering model discussed above. For the general case of planar multistatic arrays, the object illumination can come from several positions and the recording of the scattered field is made at other positions. The planar multistatic arrangement is hence defined as an array including several transmitters and receivers, which are placed at different positions within the same plane, and that the collected data include all possible combinations of these transmitters and receivers at all frequencies of operation. The imaging of an object is thus explained as the steps conducted to correlate the available information for each position within the volume of the imaged object. For this, the phase information plays the primary role, whereas the magnitude values are less important. In order to simplify the development below, the dependency on the magnitude information is often dropped without further notices. The subscript 0, as in $k_0$ or $\lambda_0$, is also removed in the following formulations to ease readability.

The Cartesian coordinate system is selected to describe the imaging problem, as it is best suitable for planar array arrangements. The general geometrical arrangement of the multistatic imaging problem is illustrated in fig. 2.1. Without loss of generality, the planar array is assumed to be in the ($z = z_a$)-plane, with its transmitters allocating the positions of $(x_t, y_t, z_a)$ and its receivers in the positions $(x_r, y_r, z_a)$. The two apertures might be overlapping or separated, however are only constrained to be in the same plane. The imaged object is located around the coordinate origin and is described by $O(x, y, z)$. For all following illustrations and figures, the red color is assigned for the transmitters and the blue one for the receivers. Tx and Rx refer to a transmitter and a receiver, respectively.

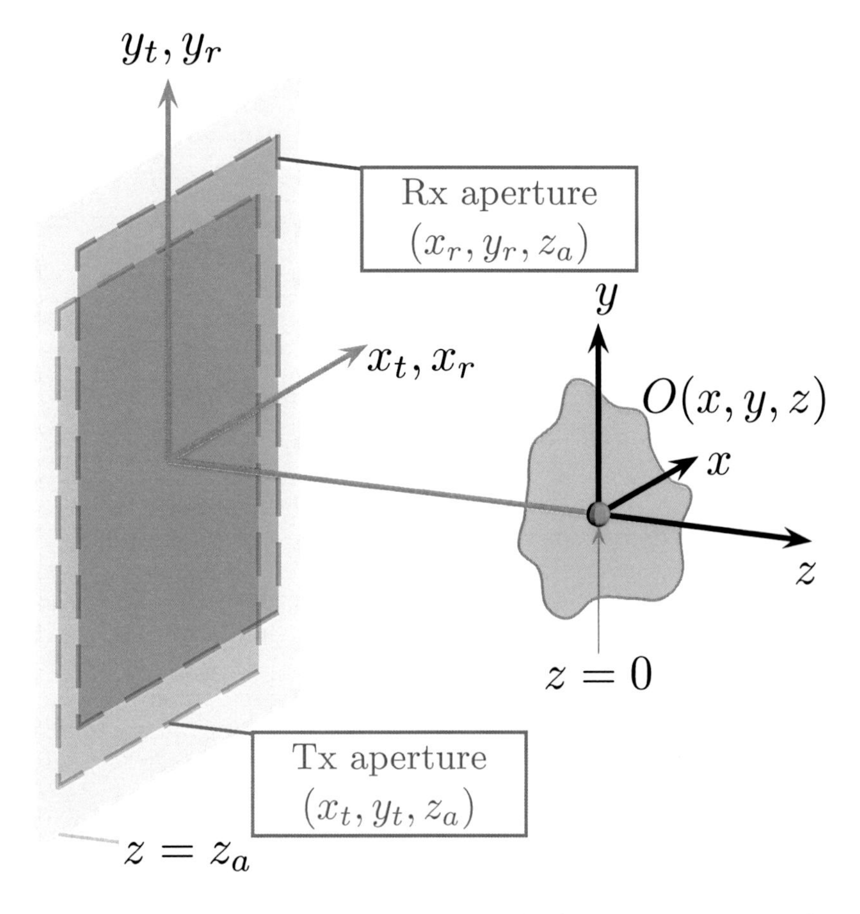


Figure 2.1: The geometrical description in the Cartesian coordinates for a general imaging problem with planar multistatic arrays. The imaged object is located around the coordinate origin, whereas the transmitters and receivers are both located in the $(z = z_a)$-plane.

The description of the scattered field in eq. 2.46 is now adapted to the notations of the imaging problem using

$$
\begin{aligned}
O(\mathbf{r}') &\Rightarrow O(x, y, z) \\
U^s(\mathbf{r}) &\Rightarrow s(x_t, y_t, x_r, y_r, k) \\
U^i(\mathbf{r}') &\Rightarrow \exp\left(-jk\sqrt{(x_t - x)^2 + (y_t - y)^2 + (z_a - z)^2}\right) \\
G(\mathbf{r} - \mathbf{r}') &\Rightarrow \exp\left(-jk\sqrt{(x_r - x)^2 + (y_r - y)^2 + (z_a - z)^2}\right) .
\end{aligned}
$$

The scattered field $s(x_t, y_t, x_r, y_r, k)$ is caused by a signal of a wavenumber $k$ emitted from a transmitter at $(x_t, y_t, z_a)$ and recorded by a receiver at $(x_r, y_r, z_a)$. The incident field $U^i(\mathbf{r}')$ corresponds to a single transmitter simplified as an isotropic radiator. The magnitude factors are dropped as discussed above. Hence, eq. 2.46 is now rewritten as

$$
\begin{aligned}
s(x_t, y_t, x_r, y_r, k) = &\int_x \int_y \int_z O(x, y, z) \\
&\cdot \exp\left(-jk\sqrt{(x_t - x)^2 + (y_t - y)^2 + (z_a - z)^2}\right) \qquad \text{(Tx term)} \\
&\cdot \exp\left(-jk\sqrt{(x_r - x)^2 + (y_r - y)^2 + (z_a - z)^2}\right) \qquad \text{(Rx term)} \\
&\cdot dx\, dy\, dz . \qquad (2.47)
\end{aligned}
$$

The process of reconstructing an image of the imaged object $O(x, y, z)$ can be only achieved within the available extent of information delivered by the imaging array. As a matter of fact, several object functions could yield to the same measurement set collected by the imaging array. Consequently, the reconstruction of the imaged object is ill-posed in a mathematical sense. Therefore, the result of the image reconstruction process delivers an estimate of the actual object function, which is referred to in the following as $\widehat{O}(x, y, z)$.

## 2.6 Image reconstruction

The calculation of an estimate $\widehat{O}$ for the object function $O$ is achieved mathematically by the inversion of the integration in eq. 2.47. This inversion can be applied through deconvolution or matched filtering methods

[23, 57, 65–67]. Deconvolution methods lack stability, and are additionally hard to achieve due to the space-variant behavior in the image, as discussed later in sec. 2.7.3. On the contrary, matched filtering is a very stable process and is used hereafter. The inversion of eq. 2.47 is described next using two approaches, one formulated in the space domain and another in the spatial frequency domain. The former mostly assists the understanding of the multistatic array design and the selection of its geometry, however the latter gives more insight about the imaging properties, e.g., resolution limits.

### 2.6.1 Space domain representation

The formulation in eq. 2.47 is already stated in the space domain, hence the matched filtering reconstruction is directly expressed as

$$\begin{aligned}\widehat{O}(x,y,z) = & \int\limits_{k}\int\limits_{y_r}\int\limits_{x_r}\int\limits_{y_t}\int\limits_{x_t} s(x_t,y_t,x_r,y_r,k) \\ & \cdot \left(\exp\left(-jk\sqrt{(x_t-x)^2+(y_t-y)^2+(z_a-z)^2}\right)\right)^* \quad \text{(Tx focusing)} \\ & \cdot \left(\exp\left(-jk\sqrt{(x_r-x)^2+(y_r-y)^2+(z_a-z)^2}\right)\right)^* \quad \text{(Rx focusing)} \\ & \cdot dx_t\, dy_t\, dx_r\, dy_r\, dk\,, \end{aligned} \tag{2.48}$$

where $(\,)^*$ indicates complex conjugate operation.

Taking into account the fact that in imaging arrays the field is transmitted and received from discrete positions, the integrations are replaced by summations and hence eq. 2.48 is rewritten as

$$\begin{aligned}\widehat{O}(x,y,z) = & \sum_{\forall k}\sum_{\forall y_r}\sum_{\forall x_r}\sum_{\forall y_t}\sum_{\forall x_t} s(x_t,y_t,x_r,y_r,k) \\ & \cdot \exp\left(+jk\sqrt{(x_t-x)^2+(y_t-y)^2+(z_a-z)^2}\right) \\ & \cdot \exp\left(+jk\sqrt{(x_r-x)^2+(y_r-y)^2+(z_a-z)^2}\right)\,. \end{aligned} \tag{2.49}$$

The physical interpretation of eq. 2.49 can be easily conducted from the assumption made by Born. In this quintuple summation, a correlation at each voxel position of the imaged object takes place by adding all scattered signals made by this voxel together. While adding the signals, the travel time of the signals is compensated on the transmission side as well as on the reception one. Thus, the contributions made due to this target voxel add constructively, whereas other voxels in the far neighborhood add destructively. This yields to a separation of information regarding the reflectivity of this voxel from the rest. The ability of the imaging system to correctly isolate the information of each target voxel from its surroundings, is described by the point spread function of the system, which is discussed later in sec. 2.7.3.

This space domain representation of the inversion problem is also known as the Back-propagation [68] or the Back-projection algorithm [69, 70]. It is similarly applied in several other imaging techniques, where the usage of scalar fields with linearized scattering problem holds true.

### 2.6.2 Spatial frequency domain representation

The scattering eq. 2.47 is now converted to the spatial frequency domain by taking the four-dimensional Fourier transform of both sides with respect to $x_t$, $y_t$, $x_r$, and $y_r$. The following identities are taken into account

$$S(k_{x_t}, k_{y_t}, k_{x_r}, k_{y_r}, k) = \mathcal{F}_{\text{4D}}\left\{s(x_t, y_t, x_r, y_r, k)\right\} \tag{2.50}$$

$$s(x_t, y_t, x_r, y_r, k) = \mathcal{F}_{\text{4D}}^{-1}\left\{S(k_{x_t}, k_{y_t}, k_{x_r}, k_{y_r}, k)\right\}. \tag{2.51}$$

Eq. 2.47 is hence transformed to

$$S(k_{x_t}, k_{y_t}, k_{x_r}, k_{y_r}, k) = \int_x \int_y \int_z O(x, y, z) \cdot \mathcal{F}_{\text{4D}}\left\{\text{Tx term} \times \text{Rx term}\right\} \cdot dx\, dy\, dz\ . \tag{2.52}$$

The Fourier transforms of the Cartesian description of the Tx and Rx terms are directly separable, namely

$$\mathcal{F}_{\text{4D}}\left\{\text{Tx term} \times \text{Rx term}\right\} = \mathcal{F}_{\text{2D}}\left\{\text{Tx term}\right\} \times \mathcal{F}_{\text{2D}}\left\{\text{Rx term}\right\}\ . \tag{2.53}$$

These two Fourier transforms can be derived using asymptotic integration methods and are proven in Appendix C to be equal to

$$\mathcal{F}_{\text{2D}}\left\{\text{Tx term}\right\} = \exp\left(j\sqrt{k^2 - k_{x_t}^2 - k_{y_t}^2}\,(z_a - z) - jk_{x_t}x - jk_{y_t}y\right),$$

and

$$\mathcal{F}_{\text{2D}}\left\{\text{Rx term}\right\} = \exp\left(j\sqrt{k^2 - k_{x_r}^2 - k_{y_r}^2}\,(z_a - z) - jk_{x_r}x - jk_{y_r}y\right).$$

By substituting these two Fourier transforms into eq. 2.53 and then into eq. 2.52 with some algebraic arrangements, it yields

$$
\begin{aligned}
S = & \exp\left(+j\left(\sqrt{k^2 - k_{x_t}^2 - k_{y_t}^2} + \sqrt{k^2 - k_{x_r}^2 - k_{y_r}^2}\right) \cdot z_a\right) \\
& \cdot \int_x \int_y \int_z O(x, y, z) \\
& \cdot \exp\left(-j(k_{x_t} + k_{x_r})x\right) \cdot \exp\left(-j(k_{y_t} + k_{y_r})y\right) \\
& \cdot \exp\left(-j\left(\sqrt{k^2 - k_{x_t}^2 - k_{y_t}^2} + \sqrt{k^2 - k_{x_r}^2 - k_{y_r}^2}\right) z\right) \\
& \cdot dx\, dy\, dz\ .
\end{aligned}
\tag{2.54}
$$

The integration in the right hand side of eq. 2.54 represents a three-dimensional Fourier transform of the object function. The components $k_x$, $k_y$, and $k_z$ of the spatial frequency domain for the object function are thus selected to be

$$k_x = k_{x_t} + k_{x_r}\ , \tag{2.55}$$

$$k_y = k_{y_t} + k_{y_r}\ , \text{ and} \tag{2.56}$$

$$k_z = \sqrt{k^2 - k_{x_t}^2 - k_{y_t}^2} + \sqrt{k^2 - k_{x_r}^2 - k_{y_r}^2}\ . \tag{2.57}$$

Using the following identities

$$\mathcal{O}(k_x, k_y, k_z) = \mathcal{F}_{\text{3D}}\{O(x, y, z)\} \quad \text{and} \tag{2.58}$$

$$O(x, y, z) = \mathcal{F}_{\text{3D}}^{-1}\{\mathcal{O}(k_x, k_y, k_z)\}\ , \tag{2.59}$$

then eq. 2.54 is rewritten to

$$
\begin{aligned}
& S \cdot \exp\left(-j\left(\sqrt{k^2 - k_{x_t}^2 - k_{y_t}^2} + \sqrt{k^2 - k_{x_r}^2 - k_{y_r}^2}\right) \cdot z_a\right) \\
& = \mathcal{O}\left(\underbrace{k_{x_t} + k_{x_r}}_{k_x}, \underbrace{k_{y_t} + k_{y_r}}_{k_y}, \underbrace{\sqrt{k^2 - k_{x_t}^2 - k_{y_t}^2} + \sqrt{k^2 - k_{x_r}^2 - k_{y_r}^2}}_{k_z}\right)
\end{aligned}
\tag{2.60}
$$

or equivalently

$$\mathcal{O}(k_x, k_y, k_z) = S \cdot \exp(-jk_z z_a)\,. \tag{2.61}$$

The extent of $S$ in the spatial frequency domain is limited according to the geometry of the transmitting and receiving apertures as well as the frequency band of operation. Consequently, the spatial spectrum of the estimate $\widehat{O}$ is identically limited to the spectral extent of $S$. The reconstructed object function using spatial frequency domain representation is hence expressed as

$$\widehat{O}(x, y, z) = \mathcal{F}_{\text{3D}}^{-1} \{S \cdot \exp(-jk_z z_a)\} \, . \tag{2.62}$$

In order to implement the inversion method suggested by eq. 2.62 in practice, the Fourier transforms must be applied using discrete Fourier transforms due to the availability of $S$ at discrete positions. Efficient implementation would require a regular equidistant sampling of $S$ in each direction of the $(k_x, k_y, k_z)$-space prior to the applying of the inverse Fourier transformation of eq. 2.62. This allows the usage of fast Fourier transform (FFT) algorithms. The available sampling is, however, irregular due to the nonlinear relation of eq. 2.57. Therefore, an interpolation must be performed to $S$ in order to map its values to a regular grid. This step has a great influence on the imaging quality of multistatic arrays, which is not addressed in this context. The spatial frequency domain representation of $S$ is very useful for the determination of the image resolutions and its spatial dependencies as addressed next.

## 2.7 Image properties

The imaging capability is governed by the design of the imaging array and the frequency band of operation. In the following sections, a discussion on the main image properties delivered by planar multistatic arrays is conducted.

### 2.7.1 Spatial spectrum

The spatial spectrum of the image describes the three-dimensional extension of the image within the spatial frequency domain, namely the $(k_x, k_y, k_z)$-space. It is equivalently named in literature as angular spectrum [65], plane wave spectrum [58], Ewald sphere [71], or K-space representation [72]. All describe the same physical meaning, however with different notations. In the following, the maximum allocation of K-space caused by multistatic image arrays is explained, followed by an analysis for the actual possible support inside the K-space, which is governed by the exact geometrical arrangement of the array and the imaged object.

#### 2.7.1.1 Allocation of K-space

By substituting eq. 2.55 and eq. 2.56 into eq. 2.57 followed by squaring of the both sides, it yields

$$\left(k_x - k_{x_t}\right)^2 + \left(k_y - k_{y_t}\right)^2 + \left(k_z - \sqrt{k^2 - k_{x_t}^2 - k_{y_t}^2}\right)^2 = k^2 . \quad (2.63)$$

This describes a set of spherical surfaces in $(k_x, k_y, k_z)$-space with a constant radius of $k$ and a variable center $C_t$, which is accordingly described by

$$C_t = \left(k_{x_t}, k_{y_t}, \sqrt{k^2 - k_{x_t}^2 - k_{y_t}^2}\right) . \quad (2.64)$$

The spherical surfaces only allocate hemispheres on the greater side of $k_z$, because eq. 2.57 constrains $k_z$ values where

$$k_z \geq \sqrt{k^2 - k_{x_t}^2 - k_{y_t}^2} . \quad (2.65)$$

The loci of the centers $C_t$ lay on another hemisphere of a same radius $k$, which is centered at the origin. Again this hemisphere is on the positive side of $k_z$ axis due to the relation of eq. 2.64. Fig. 2.2 and fig. 2.3 illustrate the geometrical steps involved in drawing these hemispheres.

The allocation of K-space is axially symmetric around the $k_z$-axis. Due to the mapping scheme of the array sampled aperture into the K-space, different allocation densities are expected within the $2k$-hemisphere. This is of particular importance when discrete Fourier transforms (DFT) are used to calculate the spectrum components made by planar multistatic arrays with regular sampling. For this discrete representation, two major zones of different filling densities within the $2k$-hemisphere are identified. The front side of the hemisphere is densely filled, whereas a half torus in its rear part is sparsely filled. This half torus is cut out of a torus placed concentric to the $2k$-hemisphere, and exhibits a major and a minor radii equal to $k$. Fig. 2.4 illustrates a slice view through the three-dimensional allocation of K-space showing these two zones.

While imaging with a signal of a bandwidth, multiple values of $k$ exist. The allocation of K-space is thus expanded to include all available values of $k$. For an imaging system utilizing a stepped-frequency continuous-wave signaling, the $k$'s are available at discrete values scaled linearly with frequency. In fig. 2.5 an illustration of the K-space allocation for three equidistant angular frequencies is presented.

#### 2.7.1.2 Support in K-space

The image resolution achieved for any object placed in front of the imaging array depends on the resultant filling of K-space caused by this object. The extent of this filling is named the support in K-space [23].

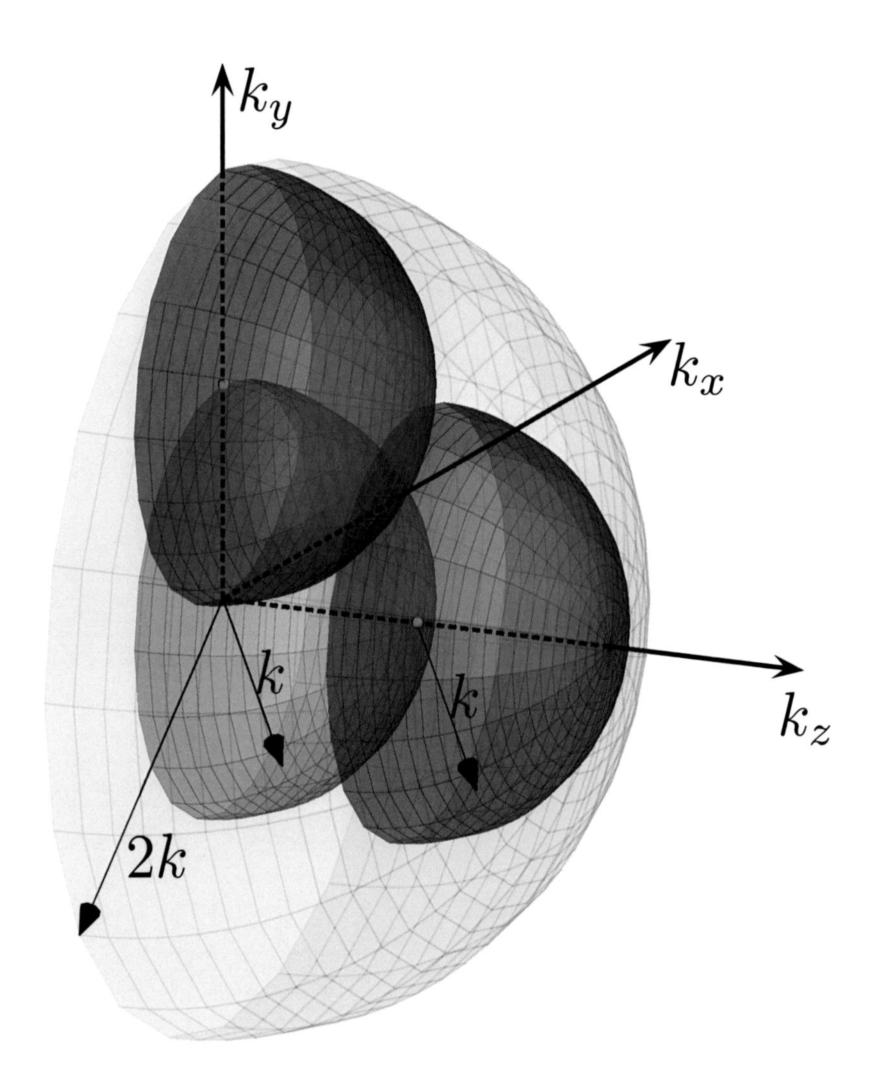


Figure 2.2: Illustration of the geometrical steps used to draw the hemispheres of the allocation of K-space. Two example hemispheres are drawn, which have their centers located on the $k$-hemisphere concentric with the coordinate origin. Their centers are thus located at $(0, 0, k)$ and $(0, k, 0)$. The outer surface of all hemispheres is another hemisphere of a radius $2k$ centered to the coordinate origin.

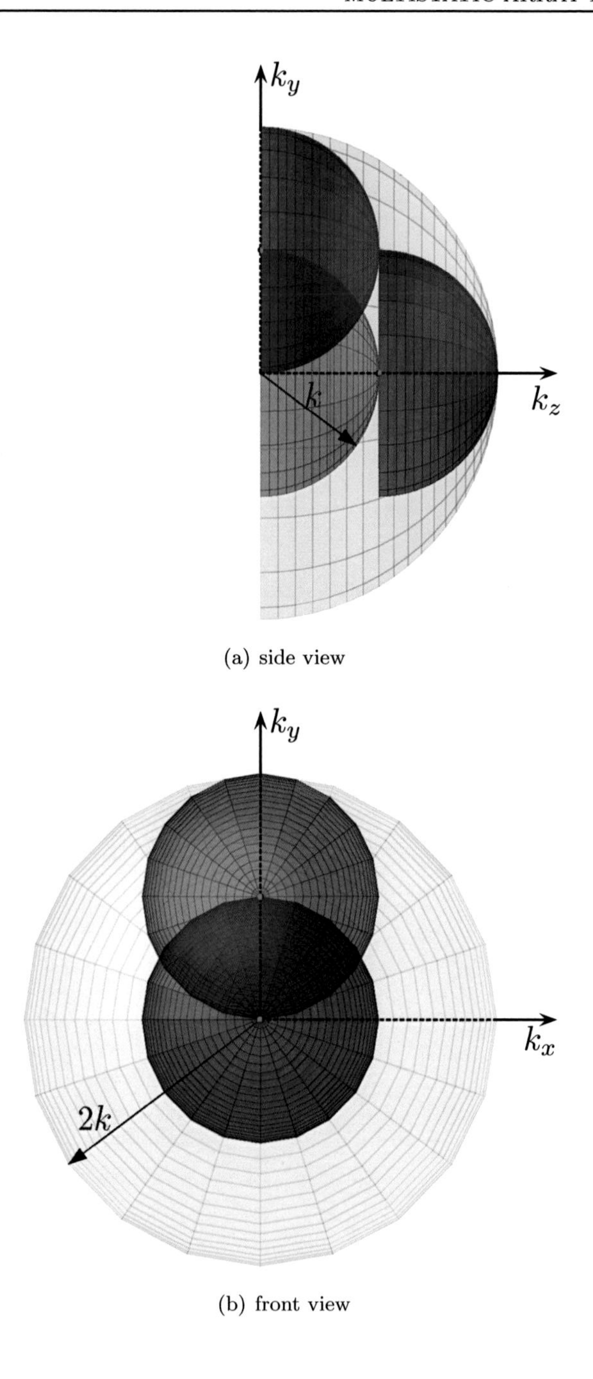


Figure 2.3: Further views of fig. 2.2 to illustrate its 3D arrangement.

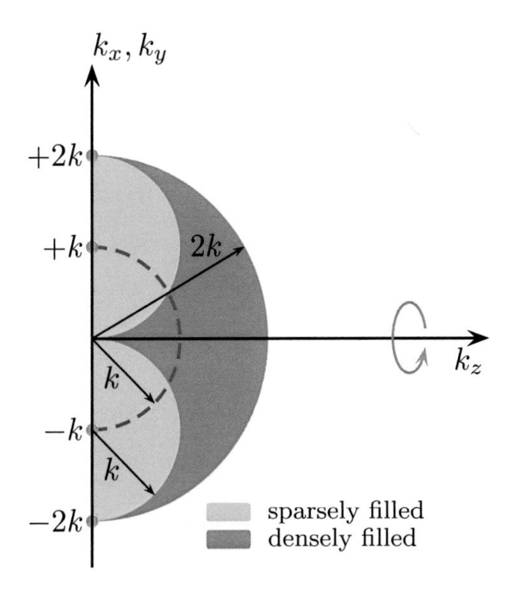


Figure 2.4: Slice view of K-space for a single value of $k$.

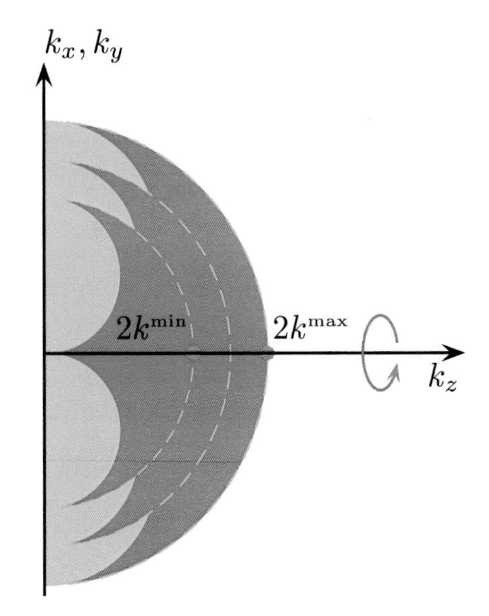


Figure 2.5: Slice view of K-space for multiple values of $k$. An example with three equidistant $k$'s is shown.

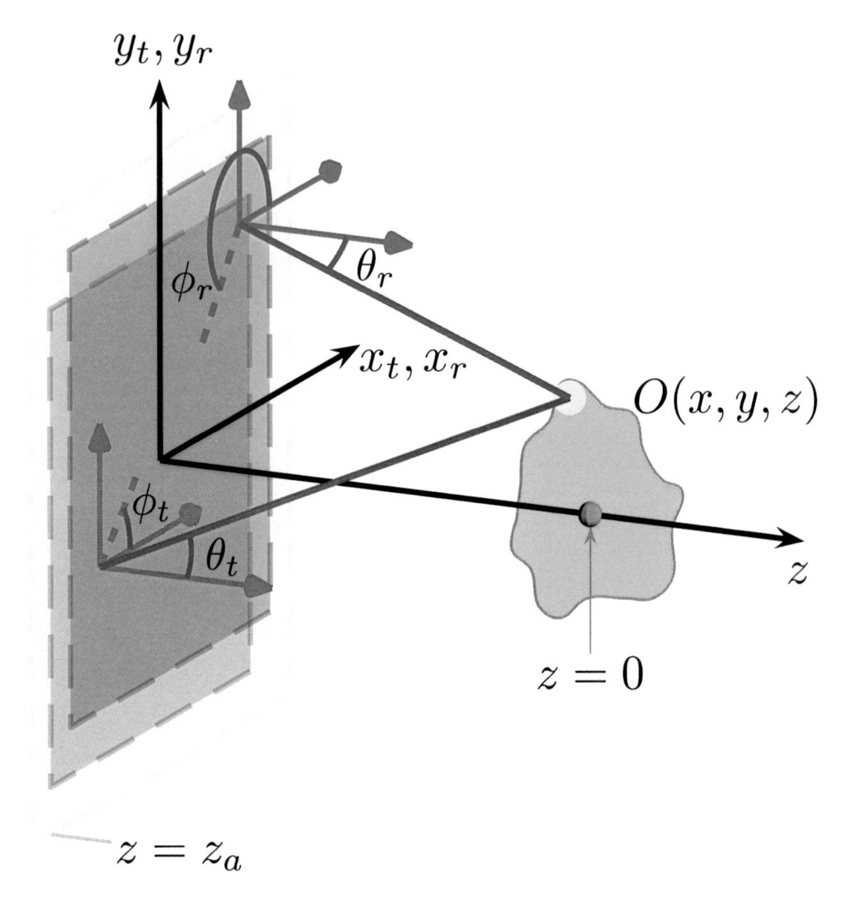


Figure 2.6: Definition of the viewing angles between the Tx and the Rx array apertures and an arbitrary point in the imaged object. The viewing angles are given by assuming spherical coordinates at the position of the Tx or the Rx antenna, which are aligned parallel to the main coordinates. The angles $\theta_t$ and $\phi_t$ describe the viewing angles with respect to the Tx antenna, whereas $\theta_r$ and $\phi_r$ with respect to the Rx antenna. Following the conventional spherical coordinates, the angles $\theta_t$ and $\theta_r$ are measured to the $z$-axis, whereas the angles $\phi_t$ and $\phi_r$ are measured from the $x$-axis to the projection of the point on the $xy$-plane.

For a general voxel at $(x, y, z)$ of the imaged object, the viewing angles relative to the Tx and Rx apertures are defined as in fig. 2.6. The angle definitions use the conventional spherical coordinate system with the origin placed at the Tx or Rx position accordingly. Following this geometrical definition, the viewing angles are thus given by

$$\theta_t = \arccos\left(\frac{z - z_a}{\sqrt{(x - x_t)^2 + (y - y_t)^2 + (z - z_a)^2}}\right), \tag{2.66a}$$

$$\phi_t = \arctan\left(\frac{y - y_t}{x - x_t}\right), \tag{2.66b}$$

$$\theta_r = \arccos\left(\frac{z - z_a}{\sqrt{(x - x_r)^2 + (y - y_r)^2 + (z - z_a)^2}}\right), \tag{2.66c}$$

$$\phi_r = \arctan\left(\frac{y - y_r}{y - y_r}\right). \tag{2.66d}$$

The values of $\theta_t$ and $\theta_r$ range from 0 to $+\pi/2$; while $\phi_t$ and $\phi_r$ range from $-\pi$ to $+\pi$. Care must be taken in evaluating equations 2.66b and 2.66d due to the sign of the arctangent function. The corresponding wavenumbers at these viewing angles are accordingly given by

$$k_{x_t} = k \cdot \sin(\theta_t)\cos(\phi_t), \tag{2.67a}$$
$$k_{y_t} = k \cdot \sin(\theta_t)\sin(\phi_t), \tag{2.67b}$$
$$k_{x_r} = k \cdot \sin(\theta_r)\cos(\phi_r), \tag{2.67c}$$
$$k_{y_r} = k \cdot \sin(\theta_r)\sin(\phi_r). \tag{2.67d}$$

And by applying eq. 2.67 to eq. 2.57, it follows

$$k_z = k \cdot (\cos(\theta_t) + \cos(\theta_r)) . \tag{2.68}$$

The limiting viewing angles define the last plane wave components recorded by the imaging array, and hence indicate the limits of the relevant support in K-space. The corresponding limiting spatial frequency components in $S$ are given, accordingly, by

$$k_{x_t}^{\min} = k^{\max} \cdot \min(\sin(\theta_t)\cos(\phi_t)), \quad (2.69a)$$
$$k_{y_t}^{\min} = k^{\max} \cdot \min(\sin(\theta_t)\sin(\phi_t)), \quad (2.69b)$$
$$k_{x_r}^{\min} = k^{\max} \cdot \min(\sin(\theta_r)\cos(\phi_r)), \quad (2.69c)$$
$$k_{y_r}^{\min} = k^{\max} \cdot \min(\sin(\theta_r)\sin(\phi_r)), \quad (2.69d)$$
$$k_{x_t}^{\max} = k^{\max} \cdot \max(\sin(\theta_t)\cos(\phi_t)), \quad (2.69e)$$
$$k_{y_t}^{\max} = k^{\max} \cdot \max(\sin(\theta_t)\sin(\phi_t)), \quad (2.69f)$$
$$k_{x_r}^{\max} = k^{\max} \cdot \max(\sin(\theta_r)\cos(\phi_r)), \quad (2.69g)$$
$$k_{y_r}^{\max} = k^{\max} \cdot \max(\sin(\theta_r)\sin(\phi_r)). \quad (2.69h)$$

The limits of the support along $k_x$- and $k_y$-directions can now be calculated from eq. 2.55 and 2.56, respectively, which yields

$$k_x^{\min} = k^{\max} \cdot \min\left(\sin(\theta_t)\cos(\phi_t) + \sin(\theta_r)\cos(\phi_r)\right), \quad (2.70a)$$
$$k_y^{\min} = k^{\max} \cdot \min\left(\sin(\theta_t)\sin(\phi_t) + \sin(\theta_r)\sin(\phi_r)\right), \quad (2.70b)$$
$$k_x^{\max} = k^{\max} \cdot \max\left(\sin(\theta_t)\cos(\phi_t) + \sin(\theta_r)\cos(\phi_r)\right), \quad (2.70c)$$
$$k_y^{\max} = k^{\max} \cdot \max\left(\sin(\theta_t)\sin(\phi_t) + \sin(\theta_r)\sin(\phi_r)\right). \quad (2.70d)$$

It is to be noticed that the limiting values in eq. 2.70 depend on $k^{\max}$, or equivalently on the highest frequency of operation. Similar to the illustration fig. 2.5, the greater $2k$-hemisphere yields the largest support along $k_x$- and $k_y$-directions. The limits of the support along $k_z$-direction are deduced from eq. 2.68 to be

$$k_z^{\min} = k^{\min} \cdot \left(\cos(\theta_t^{\max}) + \cos(\theta_r^{\max})\right) \text{ and} \quad (2.71a)$$
$$k_z^{\max} = k^{\max} \cdot \left(\cos(\theta_t^{\min}) + \cos(\theta_r^{\min})\right). \quad (2.71b)$$

The total supports along the $k_x$-, $k_y$-, and $k_z$-directions are directly given by

$$\Delta k_x = k_x^{\max} - k_x^{\min}, \quad (2.72a)$$
$$\Delta k_y = k_y^{\max} - k_y^{\min}, \quad (2.72b)$$
$$\Delta k_z = k_z^{\max} - k_z^{\min}. \quad (2.72c)$$

### 2.7.2 Spatial resolution

The resolution of an imaging system is defined in the three principle directions as the minimum geometrical distance required to separate two adjacent objects with at least 50%, i.e., 3 dB, in brightness. These distances can be directly deduced from the available signal support in K-space, which is in turn directly connected to the physical dimensions of the array transmitting and receiving apertures and the frequency band of operation. While determining these resolutions, no interference phenomenon between the two separable objects is taken into account and hence the nature of the separated objects is accordingly not considered.

The lateral resolutions $\delta_x$ and $\delta_y$ are determined by the support in $k_x$- and $k_y$-directions, respectively, whereas the range resolution by the support in $k_z$-direction. Due to the inverse Fourier transform relation of eq. 2.62, the resolutions are given approximately[§] by

$$\delta_x = 2\pi/\Delta k_x , \tag{2.73a}$$

$$\delta_y = 2\pi/\Delta k_y , \tag{2.73b}$$

$$\delta_z = 2\pi/\Delta k_z . \tag{2.73c}$$

The above equations are expressed for the very general geometrical arrangement of an imaging array. In practice, equal Tx and Rx array apertures are often used. Therefore, the resolution limits for the specific case of an array with a square aperture of side length $D$ are now considered. Two positions on a distance $L = z - z_a$ are analyzed, one in front of the array center and another one in front of the array corner, in order to compare the changes in resolution limits in dependence on position.

- Center position

  By solving eq. 2.72a and 2.72b, the lateral resolutions are calculated to be

$$\delta_x = \delta_y = \frac{c}{4f^{\max}}\sqrt{4\left(\frac{L}{D}\right)^2 + 1} , \tag{2.74}$$

  where $c$ and $f^{\max}$ denote the speed of light and the maximum frequency of operation, respectively.

[§]The used expressions ignore the signal weighting within the K-space. Weighting affects the exact shape of the corresponding focused spot, which is addressed in sec. 2.7.3 instead. The approximation is, however, sufficient for the scope of accuracy needed for the designing of an imaging system, as proved experimentally later in sec. 2.9.

The support along $k_z$-direction is given by

$$\Delta k_z = 2k^{\max} - 2k^{\min} \cdot \frac{1}{\sqrt{1+\frac{1}{2}(D/L)^2}} \ . \tag{2.75}$$

And hence the range resolution is calculated to be

$$\delta_z = \frac{c/2}{\Delta f + f^{\min}\left(1 - \frac{1}{\sqrt{1+\frac{1}{2}(D/L)^2}}\right)} \ , \tag{2.76}$$

where $\Delta f$ and $f^{\min}$ denote the signal bandwidth and the minimum frequency of operation, respectively.

The support in K-space is illustrated in fig. 2.7(a) for a single frequency, namely for $\Delta f = 0$. Next to it in fig 2.7(b), the corresponding focused spot is shown, in which the lateral extension is smaller than the range one according to the shape of the support in K-space. The spot is aligned vertical to the plane including the array aperture.

- Corner position

  Similarly, the lateral resolutions at a corner position are calculated to be

$$\delta_x = \delta_y = \frac{c}{2f^{\max}}\sqrt{\left(\frac{L}{D}\right)^2 + 1} \ . \tag{2.77}$$

The support along $k_z$-direction is given by

$$\Delta k_z = 2k^{\max} - 2k^{\min} \cdot \frac{1}{\sqrt{1+2(D/L)^2}} \ . \tag{2.78}$$

Although this support exists, it should be considered that the signal support is bent following the spherical surface in K-space. Fig. 2.7(c) illustrates the support added to K-space by a single frequency in a diagonal view, i.e., $(k_x = k_y)$-plane. Therefore, it is convenient to weight the contribution made by $f^{\min}$ for the range resolution to the half. The range resolution is consequently deduced to be

$$\delta_z = \frac{c/2}{\Delta f + f^{\min}\left(\frac{1}{2} - \frac{1}{\sqrt{4+8(D/L)^2}}\right)} \ . \tag{2.79}$$

Fig, 2.7(d) illustrates the focused spot in a diagonal view, i.e., ($x = y$)-plane. The spot is now tilt to point approximately to the array center, and has extensions corresponding to the support in K-space.

The expressions for the lateral resolutions in eq. 2.74 and eq. 2.77 are valid throughout all distances $L$ to the multistatic imaging array. The theoretical minimum for the lateral resolution is thus found from eq. 2.74 at $L \rightarrow 0$ to be

$$\delta_x^{\text{min}} = \delta_y^{\text{min}} = \frac{c}{4f^{\text{max}}} = \frac{\lambda^{\text{min}}}{4} . \tag{2.80}$$

For far-field condition, $L/D \gg 1$, thus eq. 2.74 and eq. 2.77 both converge to

$$\delta_x^{\text{far-field}} = \delta_y^{\text{far-field}} = \frac{\lambda^{\text{min}}}{2} \cdot \frac{L}{D} . \tag{2.81}$$

The dependence of the resolutions at center and corner positions on the ratio $L/D$ is illustrated along with the far-field approximation in fig. 2.8. At distances close to the imaging array, the far-field approximation is obviously invalid and the minimum resolution is limited by the $\lambda^{\text{min}}/4$ value. At far distances, the definition of the resolution is better replaced by angular resolution, which is conventionally made in optical imaging systems [73]. In practice, imaging systems designed for close range operation use $L/D$-ratios between 0.2 and 2.

Considering the result of eq. 2.76, the range resolution in far-field is given where $D/L \rightarrow 0$ and is hence calculated to be

$$\delta_z^{\text{far-field}} = \frac{c/2}{\Delta f} . \tag{2.82}$$

This is a well-known expression for classical radar systems. It is, however, the advantage of multistatic imaging at close range to enhance the range resolution much beyond the limit made by the signal bandwidth. Even without signal bandwidth, i.e., single frequency of operation, it is possible to reach range resolutions comparable to the signal wavelength. In fig. 2.9, the range resolution given by eq. 2.76 is compared to the far-field expression for a signal from 70 to 80 GHz. For instance, the range resolution at $L = D$ is 6.56 mm, while the far-field expression yields 15 mm. In practice, the enhancement of the range resolution caused by the multistatic view of the imaged object could be partially degraded by shadowing effects, however.

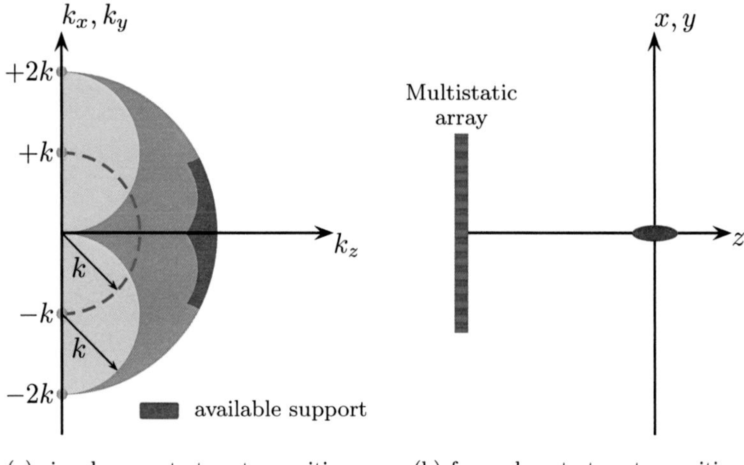


(a) signal support at center position (b) focused spot at center position

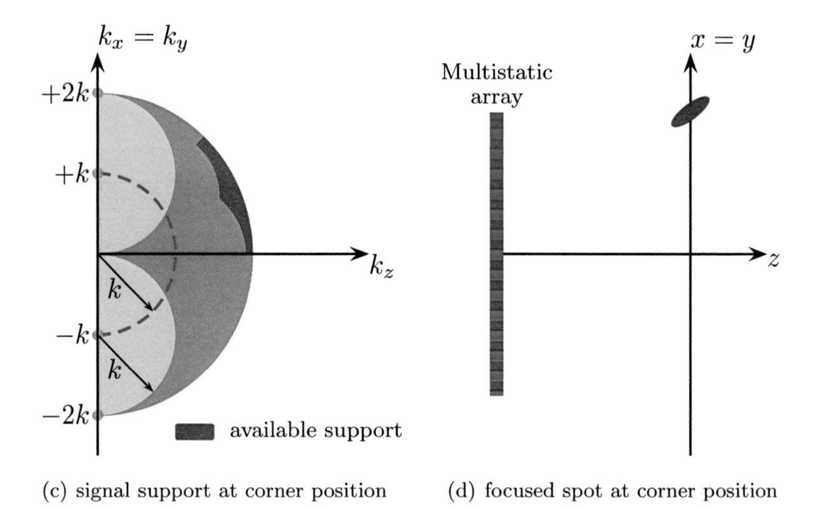


(c) signal support at corner position (d) focused spot at corner position

Figure 2.7: Left: Slice view for the support in K-space for a square array operating with a single frequency and imaging a point target at $L = D$ distance. Right: The corresponding shape of the focused spot.

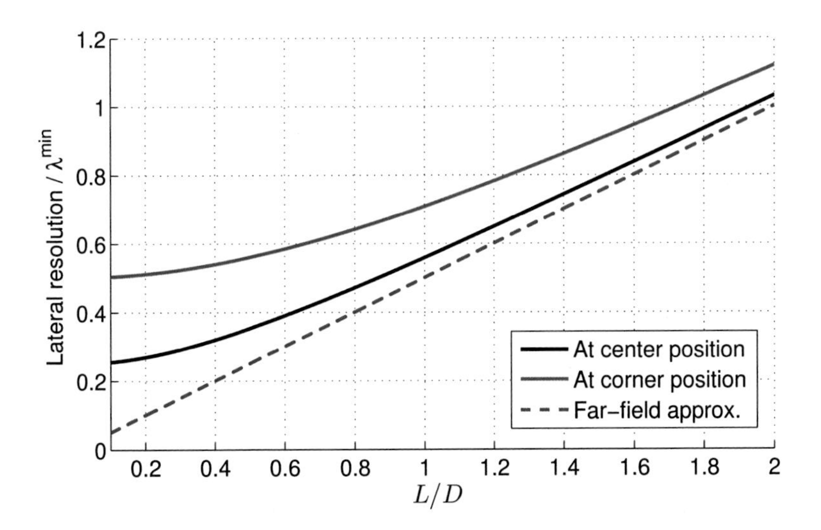


Figure 2.8: Lateral resolutions on distance $L$ using multistatic square arrays of equal transmitter and receiver apertures with side length $D$.

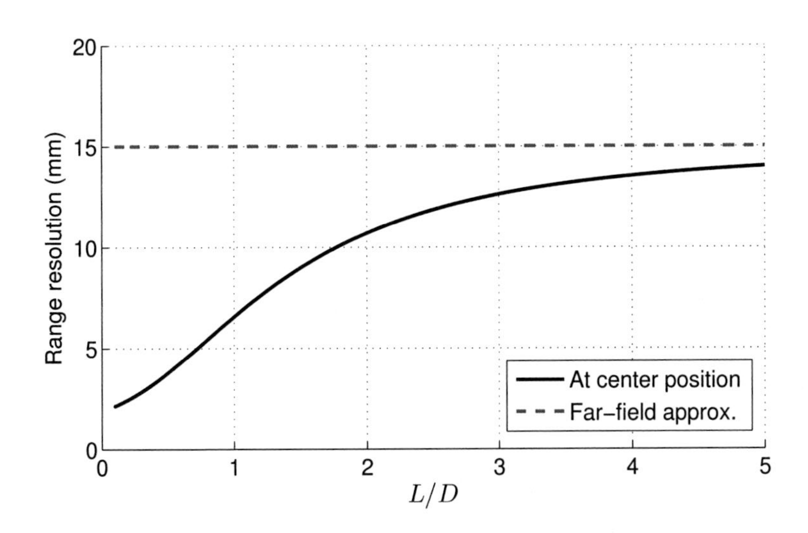


Figure 2.9: Range resolution on distance $L$ using multistatic square arrays of equal transmitter and receiver apertures with side length $D$. A frequency span from 70 to 80 GHz is used for illustration.

### 2.7.3 Point spread function

The point spread function (PSF) describes the capability of the system to image a point target. Mathematically, the point target is assumed to be a Dirac delta function, which is used to describe the collected data, and hence the inversion is done to result in a focused image at this point target. The focused spot is expected to follow the dimensions described above; it is however also important to know the signal spreading around the focused spot. To ease readability, in the following the position $(x, y, z)$ is replaced by the vector $\mathbf{r}$. Similarly, the transmitter position $(x_t, y_t, z_a)$ and the receiver position $(x_r, y_r, z_a)$ are replaced by $\mathbf{r}_t$ and $\mathbf{r}_r$, respectively.

For an assumed point target at a position $\mathbf{r}_v$, the PSF is defined as

$$\mathrm{PSF}(\mathbf{r}, \mathbf{r}_v) = \widehat{O}(\mathbf{r}) \quad \text{for} \quad O(\mathbf{r}) = \delta(\mathbf{r} - \mathbf{r}_v) \, , \tag{2.83}$$

where $\mathrm{PSF}(\mathbf{r}, \mathbf{r}_v)$ denotes the image value at position $\mathbf{r}$ due to a point target at $\mathbf{r}_v$.

Using the space domain reconstruction eq. 2.49 and considering $s^{\mathrm{PSF}}$ to be the data collected by the imaging array, the PSF is expressed as

$$\mathrm{PSF}(\mathbf{r}, \mathbf{r}_v) = \sum_{\forall k} \sum_{\forall \mathbf{r}_t} \sum_{\forall \mathbf{r}_r} s^{\mathrm{PSF}} \cdot \exp\left(+jk\left(\|\mathbf{r}_t - \mathbf{r}\| + \|\mathbf{r} - \mathbf{r}_r\|\right)\right) \, . \tag{2.84}$$

Recalling the assumption made on the transmit and receive antennas to be isotropic, the scattered field of a point target is calculated from eq. 2.46 to be

$$\begin{aligned} s^{\mathrm{PSF}} &= G(\mathbf{r}_v - \mathbf{r}_t) \cdot G(\mathbf{r}_r - \mathbf{r}_v) \\ &= \frac{e^{-jk\|\mathbf{r}_v - \mathbf{r}_t\|}}{4\pi \|\mathbf{r}_v - \mathbf{r}_t\|} \cdot \frac{e^{-jk\|\mathbf{r}_r - \mathbf{r}_v\|}}{4\pi \|\mathbf{r}_r - \mathbf{r}_v\|} \, . \end{aligned} \tag{2.85}$$

The PSF is preferably normalized, as the interest is in the shape of the function itself instead of its absolute value. When imaging large volumes, the variation of the magnitude factor in eq. 2.85 might be relevant. For the sake of simplicity, the magnitude factor is dropped hereafter. By applying eq. 2.85 in eq. 2.84, the normalized PSF is calculated to be

$$\mathrm{PSF}(\mathbf{r},\mathbf{r}_v) = \frac{1}{\sum\limits_{\forall k}\sum\limits_{\forall \mathbf{r}_t}\sum\limits_{\forall \mathbf{r}_r}(1)} \cdot \sum_{\forall k} \Bigg( \underbrace{\sum_{\forall \mathbf{r}_t} \exp\left(+jk\left(\|\mathbf{r}-\mathbf{r}_t\| - \|\mathbf{r}_v-\mathbf{r}_t\|\right)\right)}_{\mathrm{PSF_{Tx}}} \times \underbrace{\sum_{\forall \mathbf{r}_r} \exp\left(+jk\left(\|\mathbf{r}-\mathbf{r}_r\| - \|\mathbf{r}_v-\mathbf{r}_r\|\right)\right)}_{\mathrm{PSF_{Rx}}} \Bigg) . \tag{2.86}$$

The inner summations of the PSF are split into two functions, namely the $\mathrm{PSF_{Tx}}$ and the $\mathrm{PSF_{Rx}}$. These physically describe the image made for the point target due to the focusing of the transmitter and the receiver arrays separately. Due to the separability of these two functions, it is noticeable that the Tx and Rx positions can be exchanged without influence on the total focusing capability.

According to eq. 2.86, the PSF is a strong space-variant function, hence generally

$$\mathrm{PSF}(\mathbf{r},\mathbf{r}_v) \neq \mathrm{PSF}(\mathbf{r}-\mathbf{r}_v) . \tag{2.87}$$

Taking into account that the PSF is the response of the imaging system to a Dirac delta function, and the linear nature of the reconstruction described in eq. 2.49, it is hence evident that the total image reconstructed for an arbitrary object is the accumulated result of all responses to the available Dirac delta functions within the imaged volume. Accordingly, assuming that the object function is expressible as

$$O(\mathbf{r}) = \sum_{\forall \mathbf{r}_v} O(\mathbf{r}_v) \cdot \delta(\mathbf{r}-\mathbf{r}_v) , \tag{2.88}$$

then the reconstructed image $\widehat{O}(\mathbf{r})$ is alternatively described by

$$\widehat{O}(\mathbf{r}) = \sum_{\forall \mathbf{r}_v} O(\mathbf{r}_v) \cdot \mathrm{PSF}(\mathbf{r},\mathbf{r}_v) . \tag{2.89}$$

The expression of eq. 2.89 can be physically interpreted as a space-variant convolution of the PSF with the distribution of the scatterers. This is specifically true when the scatterers are geometrically separable as given by eq. 2.88. In many applications, this is not feasible as the imaged object is continuous by nature. However, in some cases where this condition is met, deconvolution of the PSF from the image in order to enhance the image quality is possible.

For appropriate imaging, the PSF must be sharp and focused. The spreading of the PSF causes overlapping of the information regarding each voxel, as given by eq. 2.89, and hence should be avoided.

Due to the involvement of two focusing functions, i.e., $\mathrm{PSF}_{\mathrm{Tx}}$ and $\mathrm{PSF}_{\mathrm{Rx}}$, several sections in the PSF can be identified. The main sections of the PSF are illustrated in fig. 2.10, and can be classified as follows.

- Focused spot:

  This is the volumetric extension of the region including the target voxel and its dimensions follow the relations discussed in sec. 2.7.2.

- Side-lobes:

  These are ringing spots directly close to the focused spot and are usually strongly coupled to it in their behavior. For good image quality, they are required to be minimized.

- Background lobes:

  However being mathematically side-lobes for the focused spot, the background lobes are usually of a low level and extend over a wide geometrical range. Their level plays a major role in defining the dynamic range of the focused image. The background lobes accumulate in the image due to the summation of eq. 2.89, and at the end become visible as a background noise in the focused image. Hence, it is advisable that the background lobes of the PSF be some order of magnitude less than the final background noise required in the focused image.

- Grating lobes:

  These are spots of a comparable level to the focused spot. They can falsify the focusing procedure and hence must be avoided. In practice, they are avoided by densely sampling the aperture, or by limiting the view angle of the antennas, or both. For multistatic imaging, grating lobes can be suppressed by purposely dislocating them in the $\mathrm{PSF}_{\mathrm{Tx}}$ and $\mathrm{PSF}_{\mathrm{Rx}}$.

- Residual grating lobes:

  For the case of incomplete suppression of grating lobes, residual grating lobes might exist. They are of special importance for multistatic imaging. Having a grating lobe in one of the $\mathrm{PSF}_{\mathrm{Tx}}$ or $\mathrm{PSF}_{\mathrm{Rx}}$, the multiplication in eq. 2.86 would result in a residual grating lobe. Consequently, they must be minimized for appropriate focusing.

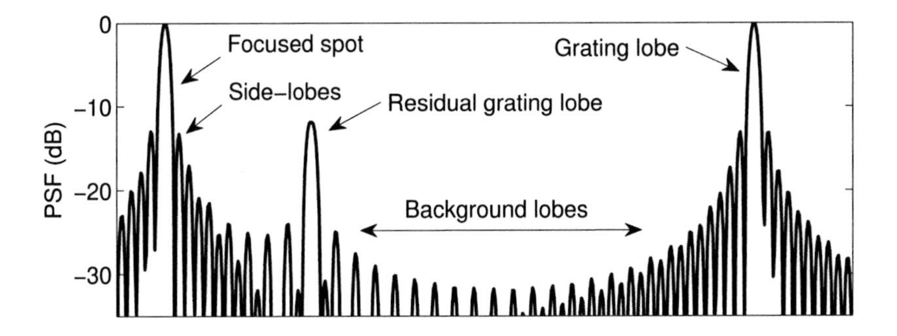


Figure 2.10: Example of a one-dimensional PSF illustrating the various sections considered in the evaluation of its quality.

## 2.7.4 Illumination and shadowing

The multistatic imaging exhibits certain characteristics regarding the illumination appearing on reflective surfaces and the development of shadows around them. For an object with smooth surfaces, the reflections become more specular than diffuse. The consequences for this phenomenon are addressed in detail in chapter 3. The multiple views with varieties of look angles offered by the multistatic arrangement cause the brightness distribution within the image to vary accordingly. Brightness correction methods might be applied to produce homogeneous illumination if necessary. On the other hand, such multiple view capability of multistatic imaging might allow the signal to reach portions of the imaged object, which are not accessible by monostatic views.

Shadowing happens when some object hinders the signal path, or when the signal is reflected away from the receiving aperture at some place. In some cases, these shadows can be beneficial to characterize an object, or to investigate its three-dimensional body. In many other cases, they are considered a drawback due to the associated loss of information about the parts in the shadow region. Generally speaking, each bistatic measurement, made with any Tx-Rx combination, includes two shadows per blocking object. In monostatic case, for instance, these two shadows coincide to a single one. Depending on the target application and the imaged object, the array arrangement should try to avoid such shadowing effects, as they reduce the viewing of the background. Such less-viewed portions will be reconstructed with less resolution, as well, due to the missing corresponding support in the K-space.

Another common reason for dark zones inside the focused image comes due to the coherent focusing of the signal at positions where strong interference patterns are present. This is a physical limitation, which is often seen when imaging semi-transparent objects of thicknesses comparable to the signal wavelength. Interference patterns thus build up and regions will appear consequently dark in the reconstructed image, where destructive interference exists.

## 2.8 Multistatic arrays

Designing a planar multistatic array exhibits a wide range of choices regarding the two-dimensional arrangement of the transmitter and the receiver apertures. Although these choices are not fully independent, it is still very challenging to find a proper array arrangement to serve a target application. In the following sections, the major design rules are discussed to enlighten about the array geometry choice; then array examples are introduced along with experimental verifications in the millimeter-wave range.

### 2.8.1 The simple design

A very simple construction of a multistatic array is shown in fig. 2.11(a), namely a plus arrangement. This geometry includes two orthogonal one-dimensional arrays, one as Tx array and another as Rx array [74]. The operation of this arrangement can be intuitively understood by considering the horizontal Tx and the vertical Rx arrays to focus the signal horizontally and vertically, respectively. Like this, a focused spot is achievable. In this example, the arrays are selected to include 128 Tx and 128 Rx elements with antenna spacing of 3.9 mm. Thus, each array is almost a half meter in length. The corresponding PSF for a central position at 0.5 m distance is shown in fig. 2.11(b) for a single frequency of 75 GHz.

### 2.8.2 Effective aperture concept

The concept of effective aperture is very advantageous for planar multistatic array designs. The effective aperture defines the far-field characteristics of the multistatic array. It does not, however, fully apply for the close range case. Fulfilling the design rules for far-field solution is considered a good approximation to start the design of the multistatic array. This approach was introduced in ultrasonic imaging systems as in [75] and [76].

Following the geometry definition in fig. 2.6, the far-field looking angles $\theta$ and $\phi$ are now described at the point $(0, 0, z_a)$ using the spherical coordinate system. The Tx aperture is named as $a_t(x, y)$, and similarly the Rx aperture as $a_r(x, y)$. The array factors [77] for the Tx array $\mathrm{AF}_{\mathrm{Tx}}$

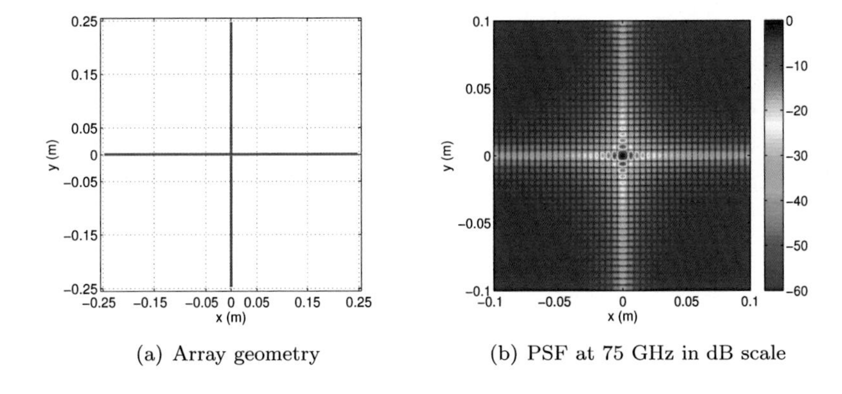


(a) Array geometry

(b) PSF at 75 GHz in dB scale

Figure 2.11: Example for a simple multistatic array of 128 Tx and 128 Rx elements arranged in a plus shape.

and for the Rx array $\mathrm{AF}_{\mathrm{Rx}}$ at some wavenumber $k$ are thus given by

$$\mathrm{AF}_{\mathrm{Tx}} = \int_{\forall x}\int_{\forall y} a_t(x,y)\, e^{-jkx(u_o-u)}\, e^{-jky(v_o-v)}\, dx\, dy\,, \text{and} \tag{2.90}$$

$$\mathrm{AF}_{\mathrm{Rx}} = \int_{\forall x}\int_{\forall y} a_r(x,y)\, e^{-jkx(u_o-u)}\, e^{-jky(v_o-v)}\, dx\, dy\,, \tag{2.91}$$

for the direction cosines $u = \sin(\theta)\cos(\phi)$ and $v = \sin(\theta)\sin(\phi)$. The $u_o$ and $v_o$ define the main beam direction. The integrals above are identified as Fourier integrals when the array factors are mapped as

$$\mathrm{AF}_{\mathrm{Tx}}(k(u_o-u), k(v_o-v)) = \mathcal{F}_{\mathrm{2D}}\left\{a_t(x,y)\right\}\,, \text{and} \tag{2.92}$$

$$\mathrm{AF}_{\mathrm{Rx}}(k(u_o-u), k(v_o-v)) = \mathcal{F}_{\mathrm{2D}}\left\{a_r(x,y)\right\}\,. \tag{2.93}$$

For general planar arrays, the two-way array factor, namely AF, is defined by the multiplication of the transmitter and the receiver array factors in space domain. Using the Fourier transform properties, the following applies.

$$\begin{aligned} \mathrm{AF} &= \mathrm{AF}_{\mathrm{Tx}} \cdot \mathrm{AF}_{\mathrm{Rx}} \\ &= \mathcal{F}_{\mathrm{2D}}\left\{a_t\right\} \cdot \mathcal{F}_{\mathrm{2D}}\left\{a_r\right\} \\ &= \mathcal{F}_{\mathrm{2D}}\{\underbrace{a_t * a_r}_{a_e}\}\,. \end{aligned} \tag{2.94}$$

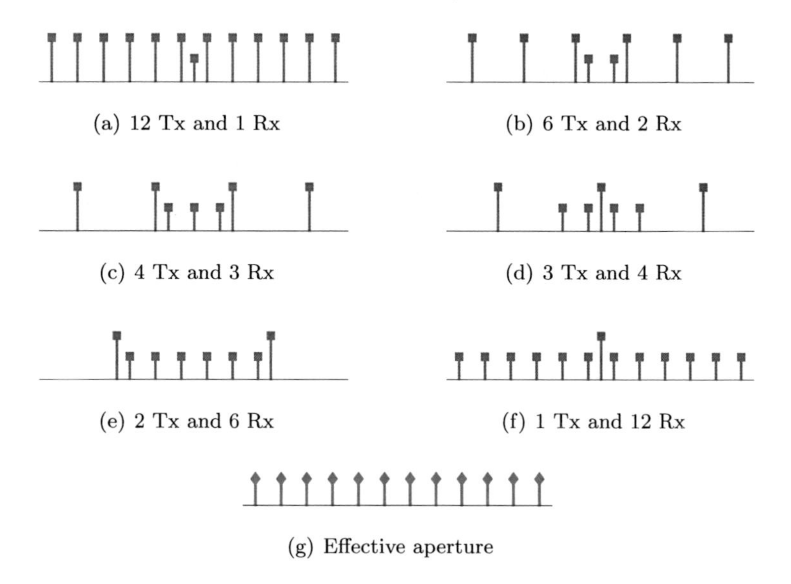

(a) 12 Tx and 1 Rx

(b) 6 Tx and 2 Rx

(c) 4 Tx and 3 Rx

(d) 3 Tx and 4 Rx

(e) 2 Tx and 6 Rx

(f) 1 Tx and 12 Rx

(g) Effective aperture

Figure 2.12: Various arrangements yielding the same effective aperture. In far-field operation, all these array arrangements are equivalent.

Eq. 2.94 states that, the two-way array factor, and hence the beam shape of the multistatic array, is fully described by an equivalent array with an aperture equal to the double spatial convolution of the Tx and Rx apertures. This aperture, named the effective aperture $a_e$, is sufficient to characterize the behavior of the multistatic array in the far-field. For the same effective aperture, the solution for the individual apertures for the Tx and Rx arrays is not unique. The total number of elements needed to allocate an effective aperture with $N_{\text{eff}}$ positions, can be minimized down to $2\sqrt{N_{\text{eff}}}$ elements, namely $\sqrt{N_{\text{eff}}}$ Tx and $\sqrt{N_{\text{eff}}}$ Rx elements. This is the case if no overlapping in the effective aperture occurs, namely no redundancy. An example for this case is shown in fig. 2.12, illustrating the various possible arrangements for the selection of the Tx and Rx arrays to yield identical effective apertures.

The idea behind the effective aperture lies in the fact that each of the single arrays, either the Tx or the Rx one, produces besides the main beam many other grating lobes. The positions of these grating lobes would perfectly match the positions of nulls in the complementary array. Thus, the two array factors exactly complement each other to produce a single effective beam and suppress all grating lobes, being either made by the Tx aperture, the Rx aperture, or both.

Moving closer to the array aperture, the far-field condition is subsequently falsified. Accordingly, the suppression of grating lobes does not take place fully, and instead residual grating lobes reside. Being close to the aperture, the grating lobes of one array spread wide and the nulls of the other array can only suppress them partially. These residual grating lobes can further be suppressed by using either a signal of a reasonable bandwidth or by increasing the redundancy in the array effective aperture. In both cases, the residual grating lobes further spread over wider angles, hence the addition of all focused signals avoids strong concentration of them at some points. For this reason, the usage of the effective aperture concept is considered as a start for an optimization problem to get the best focusing quality at the target distance to the array. This optimization is strongly coupled to the geometry of the problem and the frequency band of operation.

### 2.8.3 Design guidelines

Finding an appropriate multistatic array for a target application is a very challenging task. Here, some general guidelines and basic principles are presented, which can help in choosing an appropriate array geometry. These choices aim on one hand to ensure proper image quality from the signal processing point of view, and on the other hand to ease the realization of the imaging system later to a feasible level. It is hence clear that, subject to the available technological possibilities at the intended frequency band of operation, some of the following guidelines could be adapted or omitted.

- For any multistatic array, the positions of the transmitters and receivers are exchangeable. The two arrays would equivalently complement each other in both cases, as indicated by eq. 2.86. The choice of which array to serve as a receiver or as a transmitter one is usually made according to hardware realization reasons rather than for signal processing ones.

- For an extended support in K-space and hence for best lateral resolutions, the Tx and Rx apertures should possibly extend to the limits of the aperture in its two principle directions. This extended coverage over the aperture area is a contradicting fact with the desire to reduce the total number of elements. If redundancy-free design is desired, then the Tx and Rx elements would populate the aperture edges and be omitted in the middle. This would degrade the quality of the PSF, and hence a trade-off is to be resolved depending on the target application.

- The range resolution is specifically enhanced by using a wide frequency band of operation and by increasing the variation in $|\theta_t|$ and $|\theta_r|$, eq. 2.68. For close range imaging, the latter effect is significant, and it requires an array placement where both the aperture middle region as well as its edges are populated with Tx and Rx elements. Like this, the support in K-space along $k_z$-axis would be maximized.

- Depending on the relative position of the imaged volume and the imaging array, it should be taken into consideration to select the element placement in order to achieve possible smooth and continuous mapping to the K-space. For this, the support in K-space will be volumetrically filled and hence possible artifacts would be minimized. This is better achieved when the array topology exhibits some regular sampling and reduces possible abrupt changes in the aperture.

- Following the rules of the effective aperture discussed above, under redundant-free condition, the minimum number of elements is achieved when the total number of Tx and Rx elements are equal. Even for the case with some added redundancy, it is still beneficial to keep the total number of elements equal in order to gain more from the multistatic effect. In some applications, however, the total number of Tx and Rx elements are purposely selected to be different due to cost reasons.

- Spacing of the elements, either within the Tx or the Rx ones, must be selected to suppress the grating lobes of the whole array. For multistatic case, the grating lobes are strictly suppressed when the spacing is less than $\lambda/2$ for the shortest wavelength used, e.g., at the highest frequency of operation. This is very advantageous in comparison to monostatic case, for instance, which requires $\lambda/4$ spacing to fulfill the same condition [78]. Here, the opening angle of the used antennas can further relax this condition due to the limited angle of view they offer. It should be also noted that this suppression is required along the two principle directions, which can be either achieved simultaneously by the Tx array or the Rx array, or alternatively by the two arrays, each for a single direction.

- When selecting an array topology, it is very helpful to avoid close positions between the Tx and Rx elements. These close positions will suffer in practice from strong cross-coupling and might falsify the reflection data to some extent. Multistatic arrays offer the flexibility to avoid these geometrical arrangements, in contradiction to the monostatic imaging case, where very close positioning is essential for proper operation.

- The used antennas to realize the array elements should exhibit a stable phase center over the whole angle of view, and possibly over the frequency band of operation. Any deviation from this case would require either modification on the focusing terms used above, or otherwise cause image degradation.

- Dense placement of elements should be generally avoided due to usual practical realization constraints. In focal plane arrays, for instance, the density of elements must be increased to enhance the lateral resolution. For synthetic focusing, on the other hand, it is not needed to include dense placement to achieve better lateral resolutions. Increasing density usually yields to high thermal spots, which degrades the signal quality in practical sense. In addition, element connectivity can get complicated and antenna patterns degrade accordingly. Therefore, sparse placements are very desired.

### 2.8.4 Design examples

An imaging problem is addressed next with two possible array designs exploiting multistatic capabilities discussed above. The imaging array is to occupy an area of approximately 0.25 m$^2$, and should deliver resolutions of a few millimeters at approximately half meter distance. Following the guidelines listed above, two possible designs are investigated to demonstrate the imaging capability. The frequency of operation and hence the experimental validations are selected to be in the millimeter range, i.e., close to 75 GHz. The focusing process was made in space domain[§]. Both arrays are selected to be of a square aperture. Table 2.1 compares their details. Their geometries are illustrated in fig. 2.13. Each array consists of identical clusters, all placed on a regular grid. A detailed view of these clusters is shown in fig 2.14.

- The first multistatic array includes a total of 1024 Tx elements, distributed on $4 \times 4$ clusters, fig. 2.13(a). Each of these clusters consists of $8 \times 8$ Tx elements, as detailed in fig. 2.14(a). The receivers are distributed around each Tx cluster in 4 Rx groups per Tx cluster, where each Rx group includes $2 \times 2$ elements. Thus, a total of 256 Rx elements are distributed within the array aperture. All elements of either the Tx array or the Rx one are positioned on an equidistant grid along $x$- and $y$-directions. The unit spacing is 4 mm. The Rx elements in a Rx group are spaced to the unit spacing, and the Tx elements in a Tx cluster with double spacing, i.e., 8 mm. The Tx

[§]The numerical implementation and optimization of the focusing process is beyond the discussion presented here.

| | No. of Tx | No. of Rx | Spacing | Aperture area |
|---|---|---|---|---|
| Array 1 | 1024 | 256 | 4.0 mm | 452.0 mm × 452.0 mm |
| Array 2 | 758 | 768 | 2.9 mm | 487.2 mm × 487.2 mm |

Table 2.1: Comparison of the two investigated multistatic arrays.

clusters are placed on a grid of 128 mm step. The Rx groups are placed on a grid of 64 mm step. Accordingly, the effective aperture will result in an equidistant grid of elements continuously covering a square region without any gaps. The two arrays are then made geometrically concentric.

The PSF of the Tx and the Rx arrays are shown in fig. 2.15(a) and fig. 2.15(b), respectively. Both are calculated for a frequency of 75 GHz and are focusing centrally on 0.5 m distance. Each of the aperture functions $a_t$ and $a_r$ is of separable identical functions in $x$- and $y$-directions, e.g., array axes can be exchanged. Consequently, the $\text{PSF}_{\text{Tx}}$ and the $\text{PSF}_{\text{Rx}}$ are similarly symmetric. In both cases, the grating lobes of the individual arrays are well visible. By comparing both PSFs, any grating lobe of one array fits to a null of the other one. The PSF of the whole multistatic array is shown in fig. 2.15(c). All grating lobes are suppressed, and some residual grating lobes of low level reside. Fig. 2.17(a) shows again the three functions specifically along $(y = 0)$-line from fig. 2.15. The relative positions of the grating lobes of one array and the nulls of the other one are noticeable. Moving away from the focused spot, the grating lobes get wider and thus high residual grating lobes might develop. The redundancy included in the array arrangement helps, however, to significantly reduce the level of the residual grating lobes by widening the region of the nulls.

A verification measurement was prepared for testing the array using mechanical scanning. The mannequin shown in fig. 2.18(a) was metalized and prepared with objects. The objects were concealed behind clothes, which are not shown here. The mannequin was then placed on a half meter distance to a scanning aperture, and the reflections of the measured objects were collected by sequential mechanical scanning [51]. A frequency bandwidth from 74 to 82 GHz was used in order to reduce speckling. The magnitude of the focused three-dimensional image is afterwards projected along range direction, which is shown in fig. 2.18(b). The image is successfully focused, and the concealed objects were revealed, thus the array arrangement proves the imaging capability.

- The second multistatic array has a completely different topology, where the Tx and Rx arrays are chosen to be identical in shape, however with a rotation of 90° relative to each other, fig. 2.13(b). Again $4 \times 4$ clusters are used to construct the whole array. Each cluster includes 48 Tx and 48 Rx elements placed on the perimeter of a square, as shown in fig. 2.14(b). The total number of elements is thus 768 Tx and 768 Rx. Additionally, the unit spacing is here reduced to 2.9 mm. The Tx clusters as well as the Rx clusters are placed on a grid of 139.2 mm step, namely 48 times the unit spacing. Accordingly, the corresponding effective aperture will be again covering a square area without gaps.

  The PSFs at 75 GHz and on half meter distance from the aperture are shown in fig. 2.16(a) and fig. 2.16(b). Due to the selected aperture functions, the resultant PSFs are identical with 90° rotation. Similar as before, the grating lobes in both functions are clearly visible. These grating lobes in one PSF are co-located to the nulls in the other one. As a result, the PSF of the whole multistatic array, as shown in fig. 2.16(c), looks clear from any grating lobes. A detailed view for ($y = 0$)-line in fig. 2.16 is shown in fig. 2.17(b).

  A similar verification measurement was again prepared using the same mannequin, but instead with one concealed object behind a thick pullover, i.e., a gun. Fig. 2.18(c) shows the object setup with the pullover being pulled up for illustration. The same frequency bandwidth from 74 to 82 GHz was used. The reconstructed image is again projected along range direction in magnitude and is shown in fig. 2.18(d). Imaging capability of this multistatic array is thus also proved showing successful focusing.

The investigated two arrays above demonstrate the wide possibilities in the choice of the planar multistatic array arrangement. However being very different in geometry, similar focusing quality could be achieved, as in fig. 2.15(c) and fig. 2.16(c). Depending on the target application, one or other arrangement might be favorable. When equal number of Tx and Rx elements is desired, then the second array is more suitable. It also avoids more the dense arrangement of elements, which is of a great advantage for technological realizations. Additionally, the arrangement of the second array is easily and effectively scalable to larger apertures. Due to the linear basic structures of the Tx and Rx placement, linear increase in the number of elements results in a quadratic increase in the aperture area. This feature is very beneficial for large multistatic arrays. For these reasons, this array arrangement will be further used in the following sections as an effective and favorable arrangement for planar multistatic arrays.

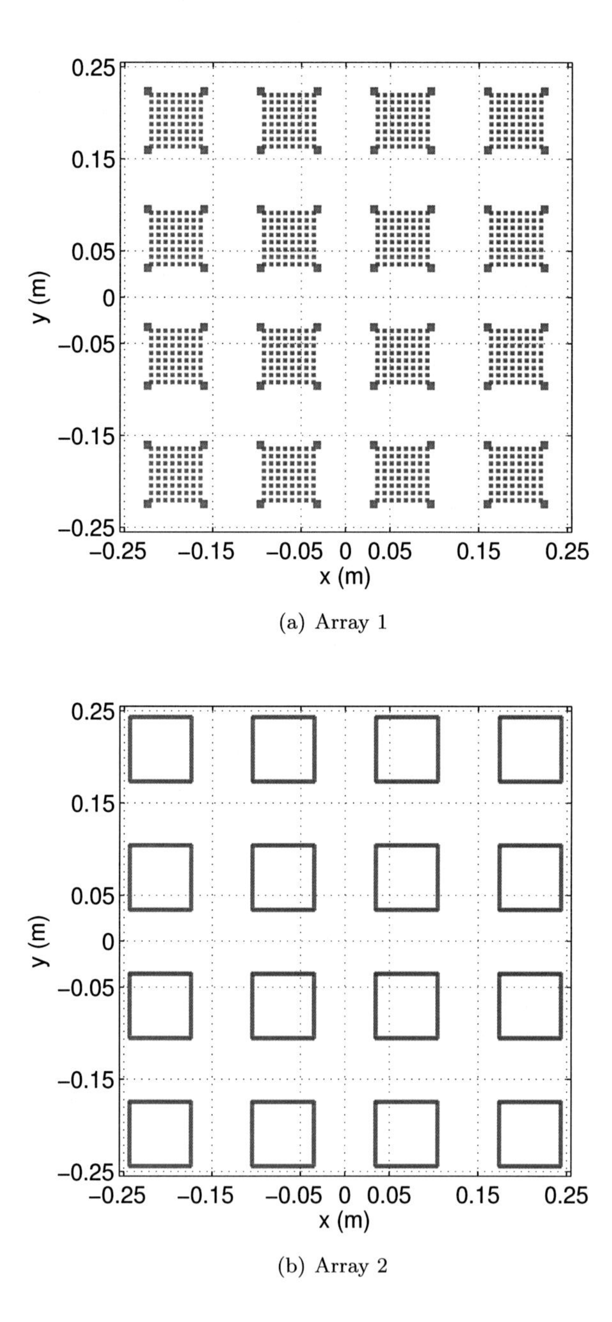


(a) Array 1

(b) Array 2

Figure 2.13: Geometry of the two investigated multistatic arrays.

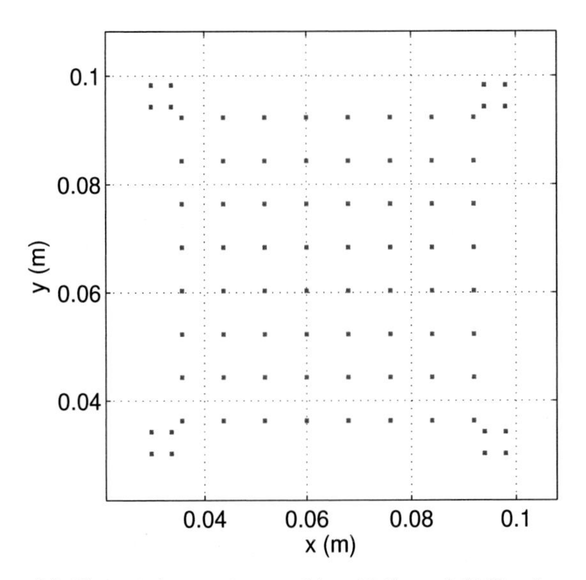


(a) Cluster of array 1 comprising 64 Tx and 16 Rx elements

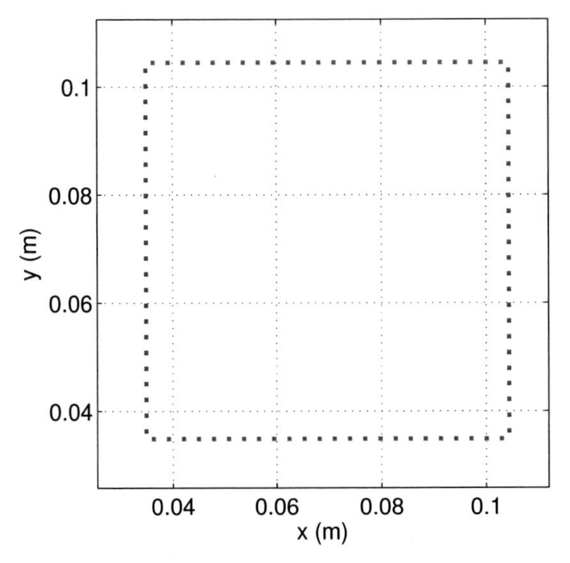


(b) Cluster of array 2 comprising 48 Tx and 48 Rx elements

Figure 2.14: Detailed illustration of the two arrays shown in fig. 2.13.

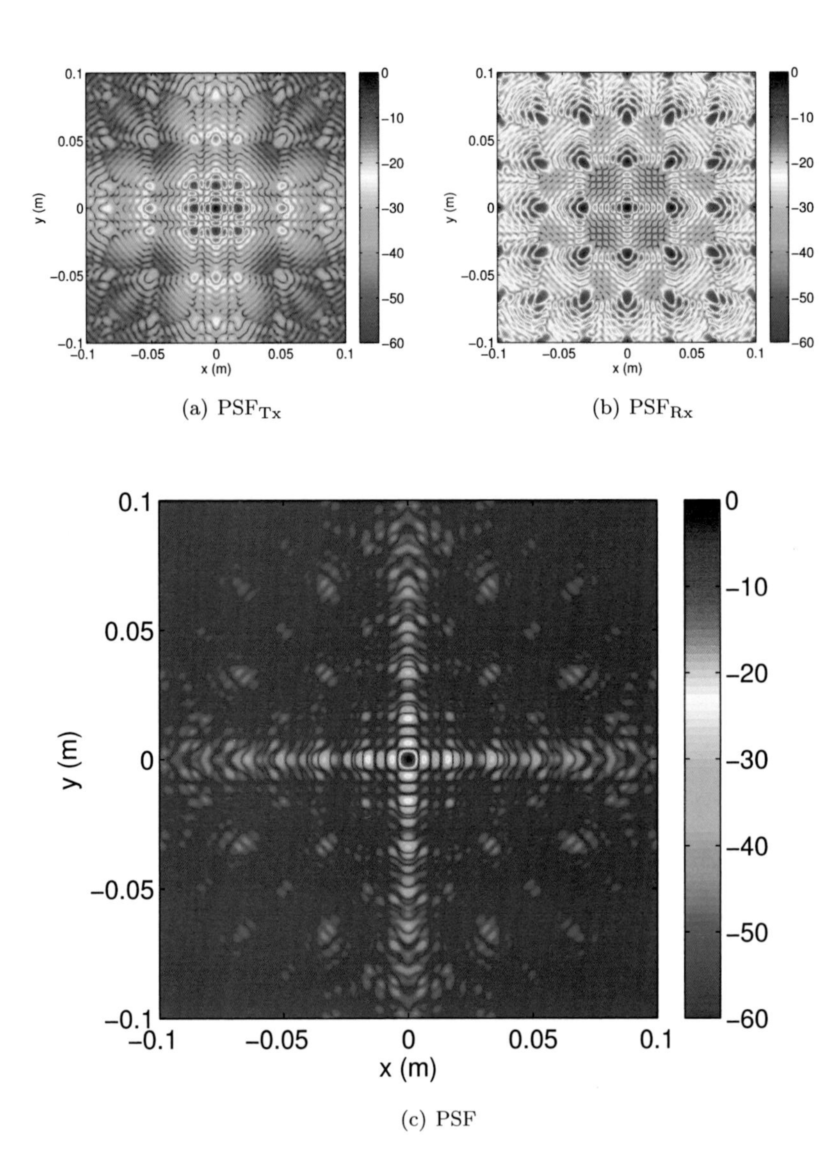


(a) $\mathrm{PSF}_{\mathrm{Tx}}$

(b) $\mathrm{PSF}_{\mathrm{Rx}}$

(c) PSF

Figure 2.15: Lateral PSF of array 1 on 0.5 m distance at 75 GHz, all shown in dB scale normalized to the maximum.

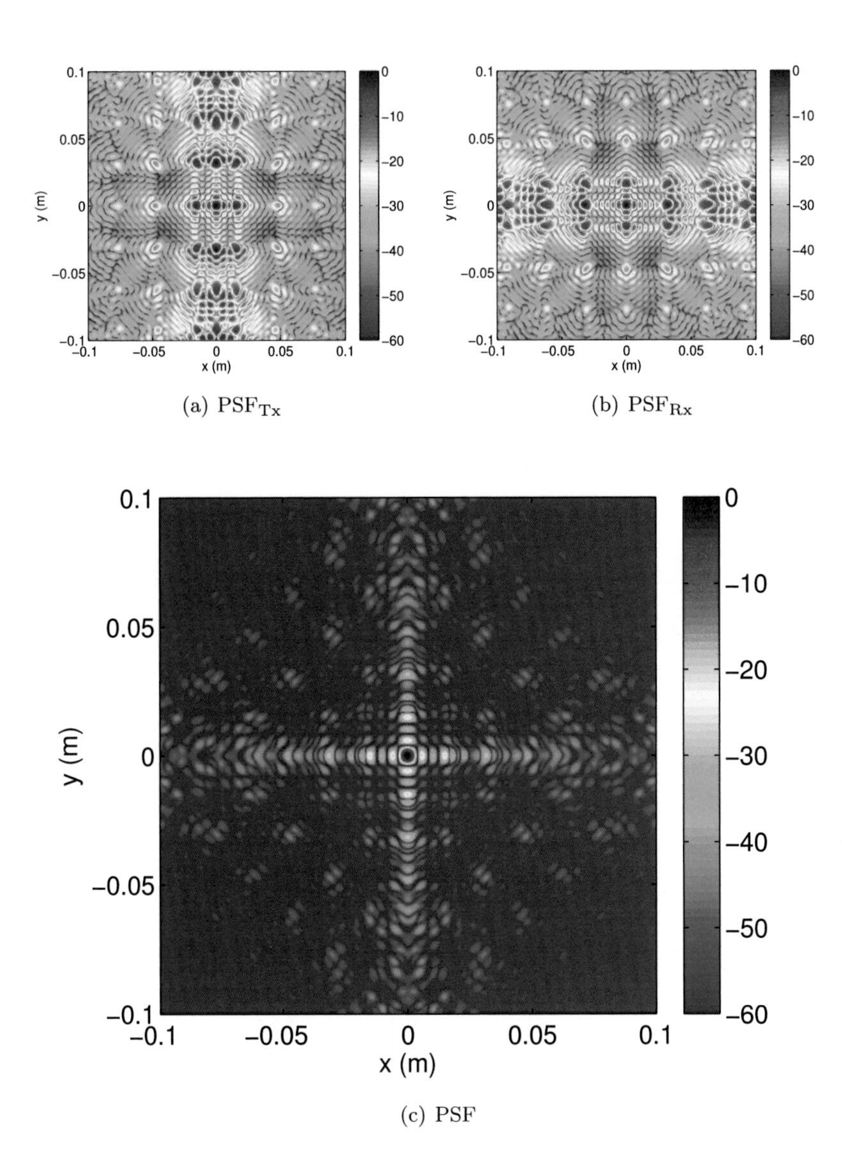


(a) $\mathrm{PSF}_{\mathrm{Tx}}$

(b) $\mathrm{PSF}_{\mathrm{Rx}}$

(c) PSF

Figure 2.16: Lateral PSF of array 2 on 0.5 m distance at 75 GHz, all shown in dB scale normalized to the maximum.

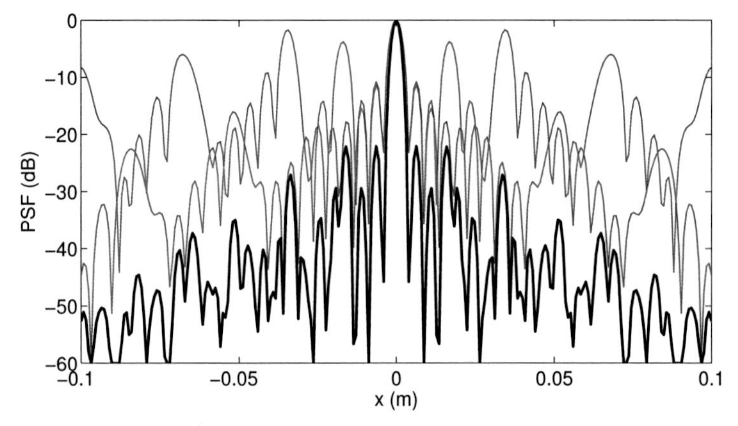


(a) Detailed view from fig. 2.15 for array 1

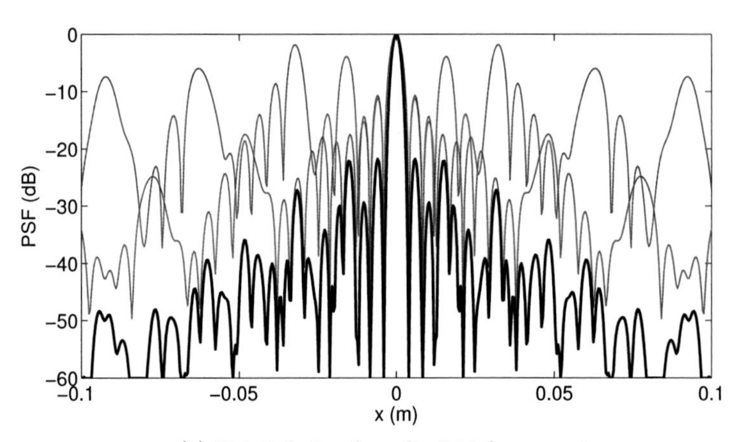


(b) Detailed view from fig. 2.16 for array 2

Figure 2.17: PSFs along ($y = 0$)-line at 75 GHz and on 0.5 m distance. Red, blue, and black lines denote the PSF of the Tx, Rx, and whole multistatic array apertures, respectively.

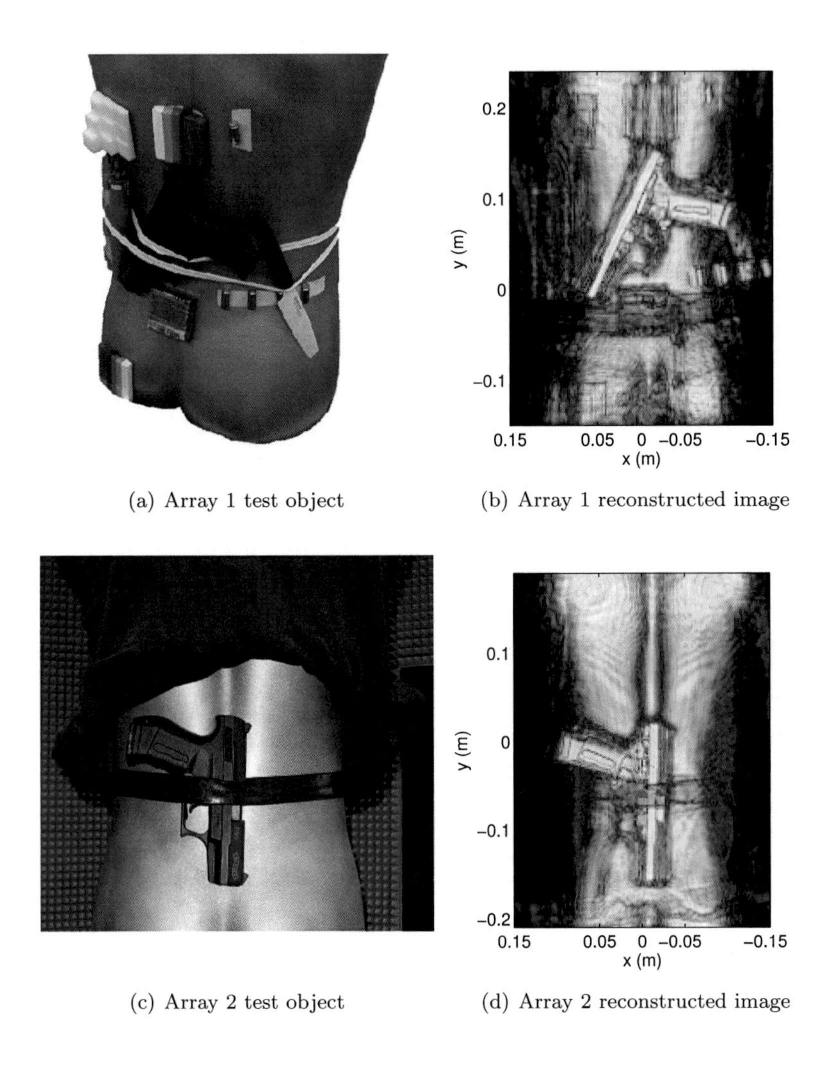


(a) Array 1 test object

(b) Array 1 reconstructed image

(c) Array 2 test object

(d) Array 2 reconstructed image

Figure 2.18: Verification measurements for the arrays shown in fig. 2.13 made with a frequency range from 74 to 82 GHz. On the left, photographs of the scanned mannequin with some concealed objects are shown. While scanning, the mannequin was prepared with clothes to hide these objects optically. The perspective view of the photograph is not identical to the view made by the scanned aperture. The images on the right side are the corresponding microwave images. They show the projections along the range direction for the magnitude of the reconstructed images. Images are presented in a logarithmic scale with 30 dB dynamic range after the equalization and smoothing introduced later in chapter 3.

## 2.9 One-square-meter array

The ultimate goal of any array synthesis problem is to find an array arrangement to satisfy certain criteria using the lowest possible number of elements. Multistatic arrays offer high reduction factor in the total number of elements relative to classical monostatic ones. Unlike the imaging systems using mechanically scanning techniques, a fully electronic solution requires a comparably reduced number of elements in order to allow its producibility. Monostatic arrays, for instance, require dense arrangement of elements with approximately $\lambda/4$ spacing. This leads to a huge number of elements in large apertures, making it very costly. Additionally, they offer a very little space for antenna integration, making it difficult to realize in fully electronic imaging.

In this section, an array of approximately one square meter aperture operating at the millimeter-wave range is introduced. The benefits of the multistatic imaging become evident for such large apertures. Its performance is then examined experimentally and compared to the theoretical expectations derived earlier in sec. 2.7.

### 2.9.1 Array geometry

The geometry used in the array introduced in fig. 2.13(b) is fully parameterizable using the antenna spacing $\Delta a$, the number of antennas per a single line of a cluster $N_a$, and the number of clusters per a side of the square aperture $N_c$. The array geometry offers a high modular and scalable aperture. The antenna spacing $\Delta a$ is better chosen to be equal or close to the value of $\lambda/2$, in order to avoid grating lobes. Fig 2.19 illustrates the scaled version of the multistatic array to reach an aperture area of approximately 1 m$^2$. Its geometry is accordingly defined by

$$N_c = 4,\ N_a = 48,\ \text{and}\ \ \Delta a = 2.9\ \text{mm}\ . \tag{2.95}$$

The array is to operate from 70 to 80 GHz, thus $\Delta a$ is equal to $0.725\lambda$ at the middle frequency.

The geometrical relations of the parameters are denoted in fig. 2.20. The total area occupied by the array is hence calculated as

$$\text{Array area} = \left((2N_\text{c} - 1) \cdot N_\text{a} \cdot \Delta a\right)^2\ . \tag{2.96}$$

For the proposed array, this yields an area of 0.95 m$^2$, hence named one-square-meter array.

The total number of Tx elements $N_t$ and Rx elements $N_r$ are equal; and are given by

$$N_t = N_r = 2N_\text{a} \cdot {N_\text{c}}^2\ , \tag{2.97}$$

thus the total number of elements in the multistatic array is equal to

$$N_{\mathrm{all}}^{\mathrm{multi}} = N_t + N_r = 4N_{\mathrm{a}} \cdot {N_{\mathrm{c}}}^2 \ . \tag{2.98}$$

The proposed array hence includes 1536 Tx and 1536 Rx elements, with a total of 3072 ones.

Next, a comparison with monostatic arrays is made. For an equivalent sampling of monostatic case, the antenna spacing must be halved, i.e., sampling with $\Delta a/2$. Thus, the filling of the same aperture area with monostatic antennas will result in a total number of elements given by

$$N_{\mathrm{all}}^{\mathrm{mono}} = (2 \cdot (2N_{\mathrm{c}} - 1) \cdot N_{\mathrm{a}} + 1)^2 \ . \tag{2.99}$$

The reduction factor of the total number of elements used by a monostatic array compared to the a multistatic one occupying the same square aperture is given by

$$\frac{N_{\mathrm{all}}^{\mathrm{mono}}}{N_{\mathrm{all}}^{\mathrm{multi}}} \approx N_a \cdot (2 - 1/N_c)^2 \ . \tag{2.100}$$

For $N_c = 1$, the reduction factor is equal to $N_a$ itself, which is the case when the antennas are all positioned on the perimeter of the aperture. The asymptotic reduction factor for $N_c \to \infty$ is $4N_a$. Hence, the gained reduction factor in practice will range between $N_a$ and $4N_a$. For instance, the proposed array offers a reduction factor of 147, which is a great help for system realization and cost reduction.

Furthermore, it is interesting to determine the ratio of the total number of measurements collected by the two arrays for the same aperture. The total number of measurements collected by the multistatic array per frequency of operation is equal to

$$N_t \cdot N_r = 4{N_{\mathrm{a}}}^2 \cdot {N_{\mathrm{c}}}^4 \ . \tag{2.101}$$

For monostatic case, each element delivers a single measurement, hence $N_{\mathrm{all}}^{\mathrm{mono}}$ is equal to the total number of measurements. By comparing both, the increase in the total number of measurements delivered by the multistatic array relative to the monostatic one is equal to

$$(2N_c - 1)^2/{N_c}^4 \ . \tag{2.102}$$

Fig. 2.21 illustrates the values made by this ratio. While increasing $N_c$, the number of measurements can rapidly increase. Therefore, in case of very large apertures, the array might be further clustered in sub-arrays in order to avoid development of huge data volumes to process during the image reconstruction. For $N_c = 4$, the ratio is equal to 5.224.

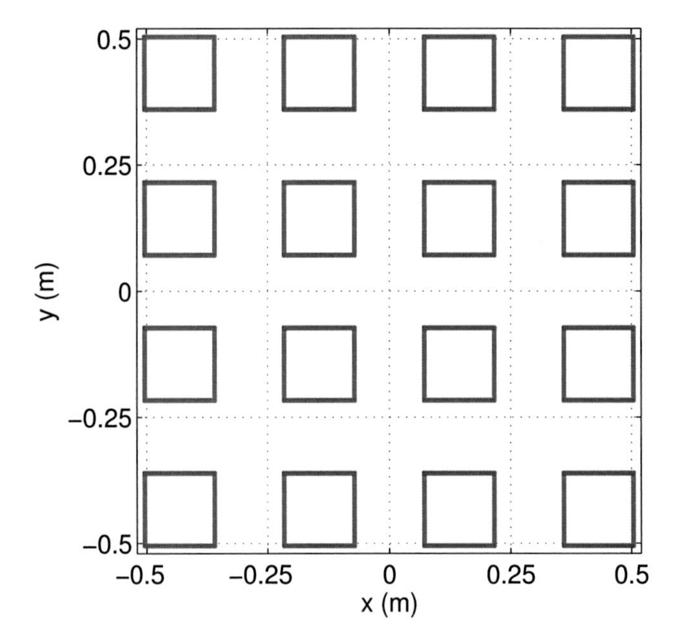


Figure 2.19: Geometry of the one-square-meter array. The aperture includes a total of 1536 Tx and 1536 Rx elements, all grouped in 16 clusters.

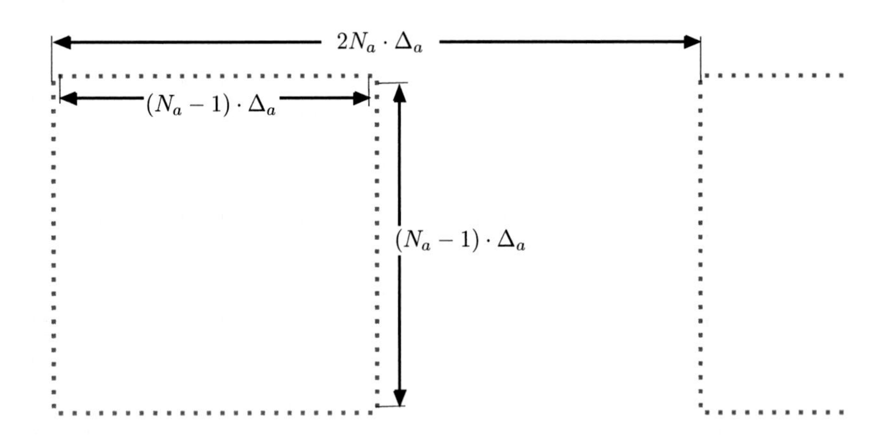


Figure 2.20: Definition of geometry parameters of the array arrangement. Illustration for the case $N_a = 24$ is shown.

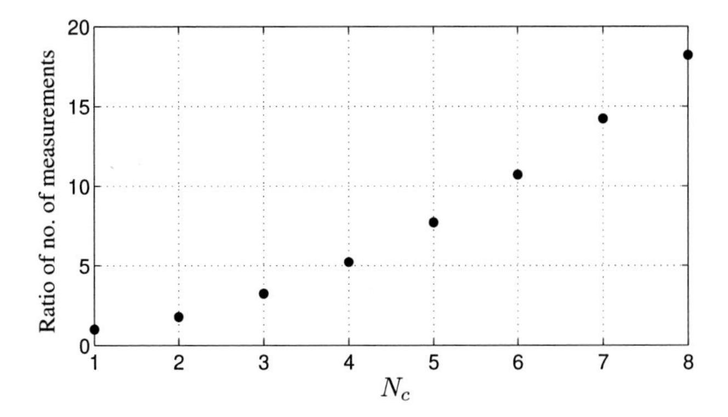


Figure 2.21: Ratio of the number of collected measurements made by a square multistatic array relative to a monostatic one occupying the same aperture. The diagram illustrates the progressive increase with $N_c$ value.

### 2.9.2 Focusing quality

The focusing is examined at two different positions, namely at center and corner positions, both at one meter distance from the aperture. These two positions are analogous to the positions considered in sec. 2.7.2. The PSFs and the corresponding supports in the spatial spectrum are simulated for a signal bandwidth of 70 to 80 GHz using 64 equidistant frequency steps. The array aperture is positioned at $z_a = -1$ m.

For the center position, the target point is at the coordinate origin. Its PSF is shown fig. 2.22 in two views, one for evaluating the lateral focusing and another for the range focusing. The corresponding support in the spacial spectrum is illustrated in fig. 2.23, in two projections along $k_z$- and $k_y$-directions. The dashed white curves show the boundaries of the $2k^{\max}$ hemisphere, which corresponds to the maximum frequency of 80 GHz. The suppression of the residual grating lobes is here noticeably enhanced compared to the single frequency case discussed in sec. 2.8.4 due to the used signal bandwidth.

For the corner position, the target point is assumed at $(0.5, 0.5, 0)$ m. Its PSF is shown in fig. 2.24, again in two views which are centered to the target point. The corresponding support in the spacial spectrum is also illustrated in fig. 2.25, in the two projections along $k_z$- and $k_y$-directions. The support is rotated inside the K-space following the angles of view relevant to the target point. The change in the dimensions and shape of the focused spot is clearly correlated to the available support in K-space. As expected, the focused spot gets narrower along the dimensions where the support is wider, and vice versa.

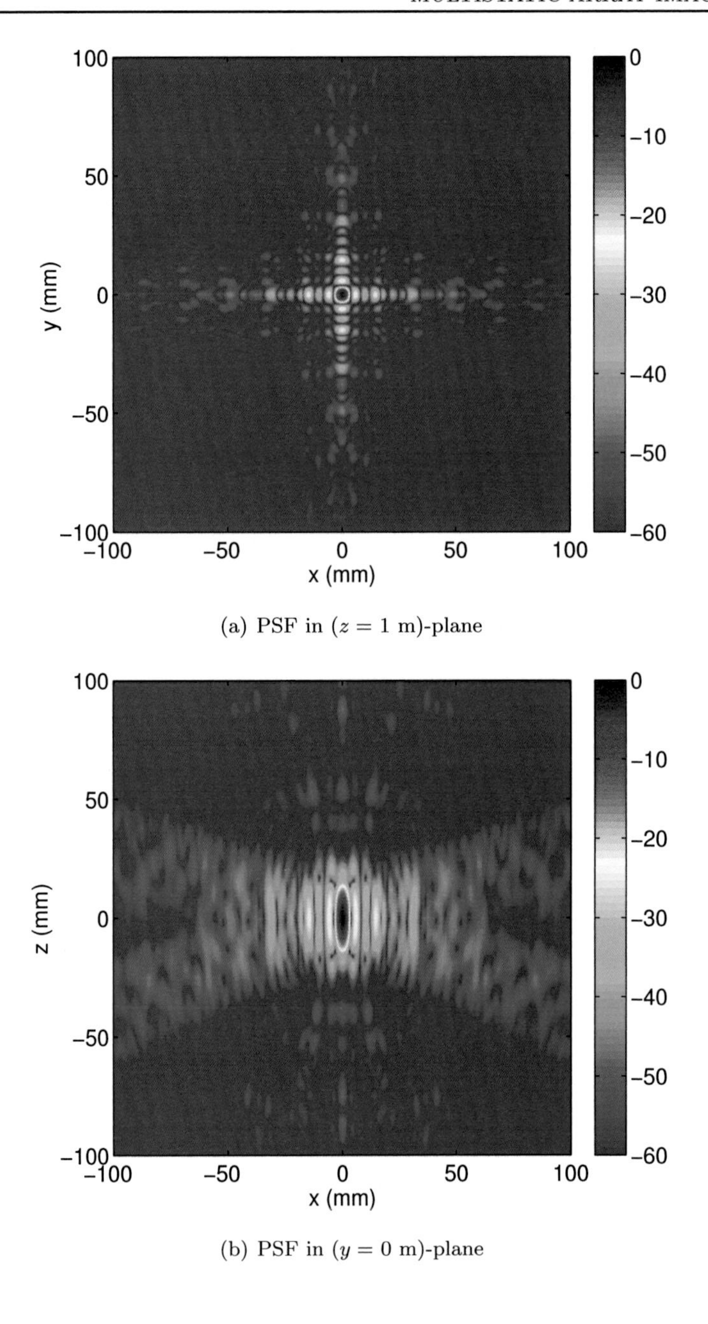


Figure 2.22: Simulation of PSF for a target at a center position on 1 m distance to a square multistatic array of 1 m$^2$ aperture operating from 70 to 80 GHz using 64 frequency steps. Levels are in a normalized dB scale.

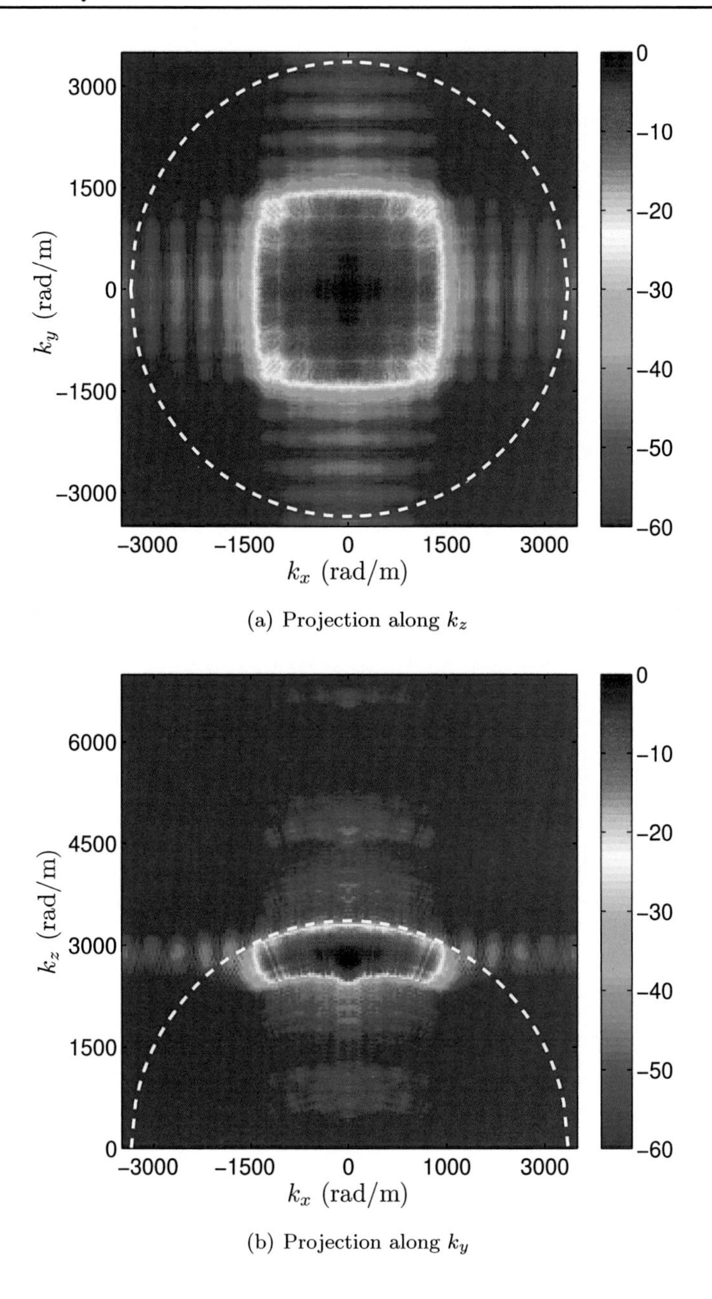


(a) Projection along $k_z$



(b) Projection along $k_y$

Figure 2.23: Signal support in spatial frequency domain of the PSF in a center position simulated in fig. 2.22. The dashed white curves show the boundaries of the $2k^{\max}$-hemisphere. Levels are in a normalized dB scale.

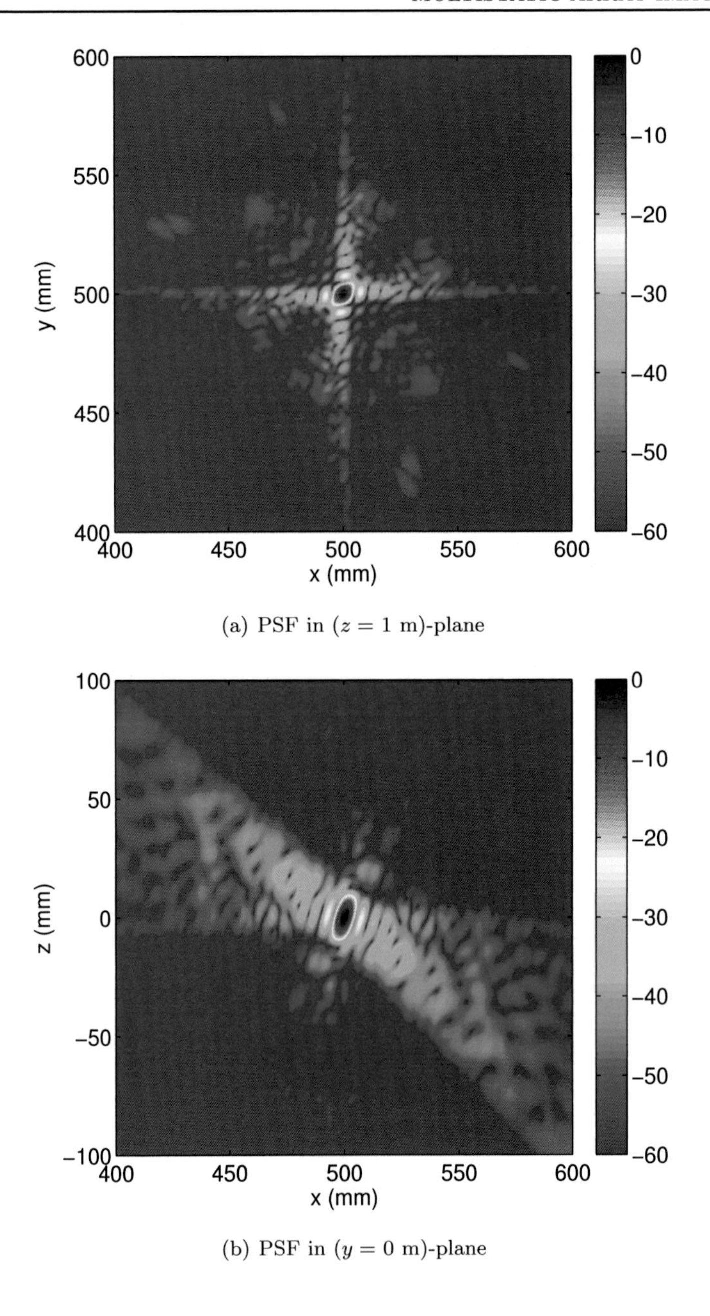


(a) PSF in ($z = 1$ m)-plane

(b) PSF in ($y = 0$ m)-plane

Figure 2.24: Simulation of PSF for a target at a corner position on 1 m distance to a square multistatic array of 1 m$^2$ aperture operating from 70 to 80 GHz using 64 frequency steps. Levels are in a normalized dB scale.

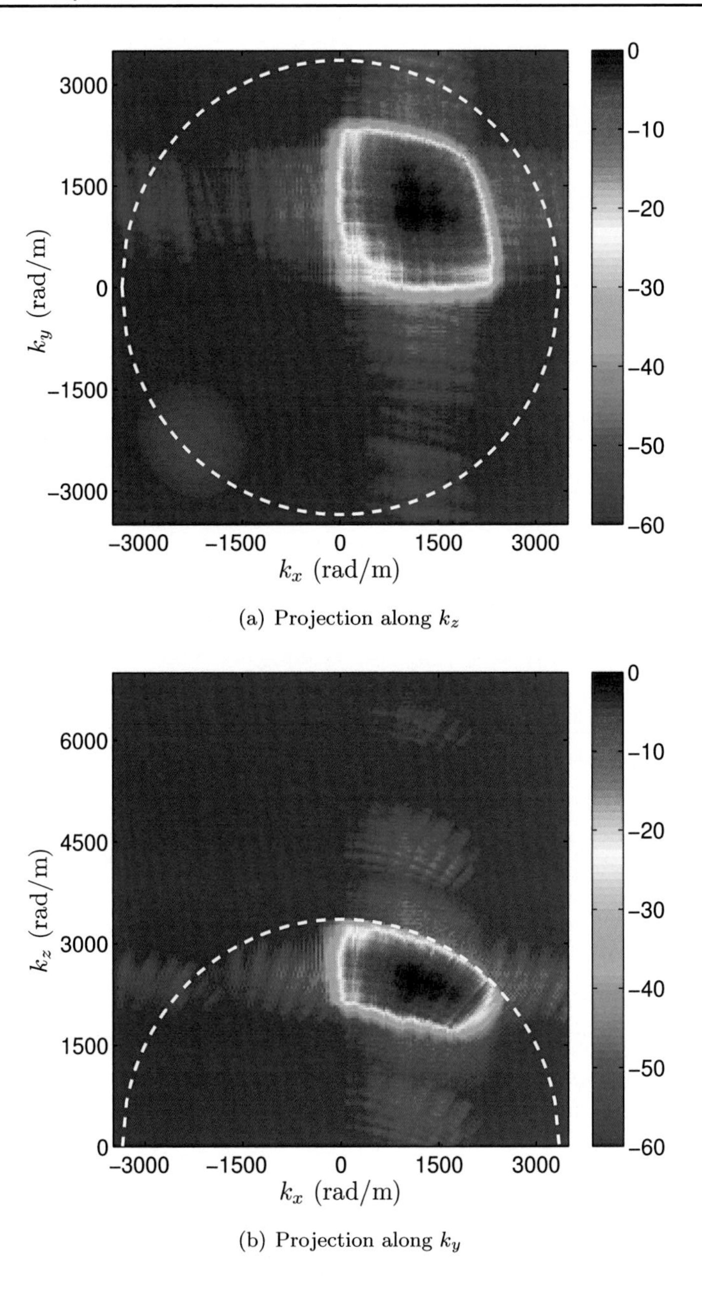


(a) Projection along $k_z$



(b) Projection along $k_y$

Figure 2.25: Signal support in spatial frequency domain of the PSF in a corner position simulated in fig. 2.24. The dashed white curves show the boundaries of the $2k^{\max}$-hemisphere. Levels are in a normalized dB scale.

### 2.9.3 Measurement verifications

The imaging capability of the array is verified experimentally using a resolution test object and a mannequin. Measurements are collected at each Tx-Rx-frequency combination, i.e., direct acquisition in frequency domain from 70 to 80 GHz in 64 steps. Afterwards, they are weighted for a smooth effective aperture following a Kaiser window (shape factor of two) as presented later in sec. 3.3. The images are focused using space domain reconstruction technique. The spatial resolutions are analyzed for the center position and at one meter distance, identical to the simulations presented above. In order to measure the lateral resolution, the lateral separation between two objects is considered. This method cannot be performed for determining the range resolution due to the signal blockage caused by the first surface. Instead, the range resolution is determined by comparing the brightness levels of a test object utilizing different range stepping, which is reconstructed at the range position of the central step. Using these brightnesses, the 3-dB-width of the PSF along range can be calculated by curve fitting them to a sinc function. Sinc function is considered a good approximation for the PSF within its focused spot, making the 3-dB-width also a close estimation for the actual resolution.

The resolution test object, shown in fig. 2.26(a), includes flat structures, e.g., Siemens star and linear slots made in a metal sheet of 2 mm thickness, and a three-dimensional (3D) metal object. The Siemens star is made of 32 alternating sectors; and the slot openings are prepared in four orientations using alternating structure with 4 mm width per slot and in an 8 mm raster. The 3D object consists of four steps with a height difference of 5 mm per step. Each step is 20 mm times 20 mm in area. Slant view is shown in fig. 2.28(a) to illustrate its construction.

The resolution test object is placed parallel to the array surface at the center position. The corresponding image is shown in fig. 2.26(b). The 4 mm slots are clearly resolved in the four orientations. The sectors of the Siemens star are resolved down to the lateral resolution limit. A detailed view of the image is shown in fig. 2.27(b), demonstrating a minimum diameter of 20 mm for the resolved part of the star. This corresponds to a lateral resolution of $20\,\pi/32$ mm, i.e, nearly 2 mm.

The range information in the 3D image is better illustrated using 3D rendering of the magnitude image. Fig. 2.28(b) shows the rendered image of the range resolution test object, and clearly visualizes the positions of the object steps. The brightnesses in the corresponding plane including the third step, counted from the bottom, are examined against the neighboring two other steps and hence curve-fitted to a sinc function. The resultant 3-dB-width of the PSF is thus 7.3 mm. Table 2.2 compares the achieved measured values with the theoretical calculations from sec. 2.7.2.

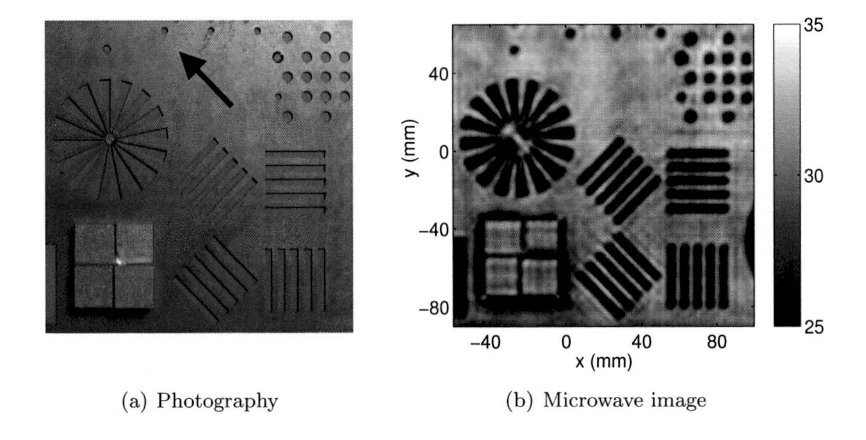


(a) Photography (b) Microwave image

Figure 2.26: Experimental verification of image resolution at 1 m distance to the one-square-meter array using a frequency range from 70 to 80 GHz. The arrow points the applied polarization. The reconstructed 3D image is projected along range direction and presented in dB scale.

| | Theoretical | Measurement |
|---|---|---|
| Lateral resolution | 2.1 mm | 2.0 mm $\pm$0.2 mm |
| Range resolution | 6.5 mm | 7.3 mm $\pm$1.0 mm |

Table 2.2: Comparison of resolution values for the center position at 1 m distance to the one-square-meter array with the expected tolerances.

The imaging quality is now verified using a large object, namely a metalized mannequin prepared with cloths and several concealed objects, fig. 2.29(a). The concealed objects are a ceramic knife, a pair of scissors, a piece of a modeling clay, and a P99 pistol model, fig. 2.29(c). The mannequin was placed at around 80 cm distance to the array. The projected magnitude image is shown in fig. 2.29(b) in logarithmic view of 35 dB range, which demonstrates high image quality being free of any significant artifacts. The range information is additionally used to color this image as in fig 2.29(d), where red to blue presents close to far distance relative to the array, respectively. Fig. 2.30(a) shows a photograph of the imaged pistol, and next to it in fig. 2.30(b) a detailed view of the reconstructed image is present. Due to the detailed resolution and the high image dynamic range offered by the imaging array, great details of its surface as well as its internal parts, e.g., magazine, are clearly visible.

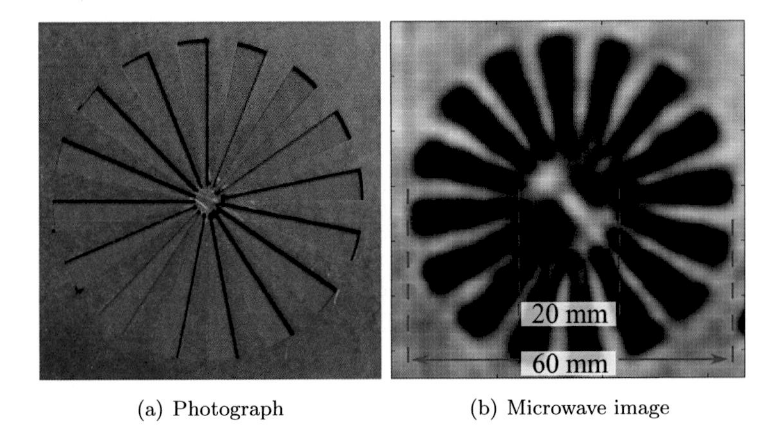


(a) Photograph (b) Microwave image

Figure 2.27: Detailed view of the Siemens star from fig. 2.26. A minimum diameter of 20 mm for the resolved part of the star corresponds to a lateral resolution of 2 mm. The shape of the unresolved part of the star indicates the effect of the used polarization.

(a) Photograph (b) Microwave image

Figure 2.28: The test object used to examine the range resolution. It consists of four steps made of bulky metal. The spacing between each two successive steps is equal to 5 mm. A tilt view of the rendered three-dimensional image is shown to demonstrate the range information in the reconstructed magnitude image.

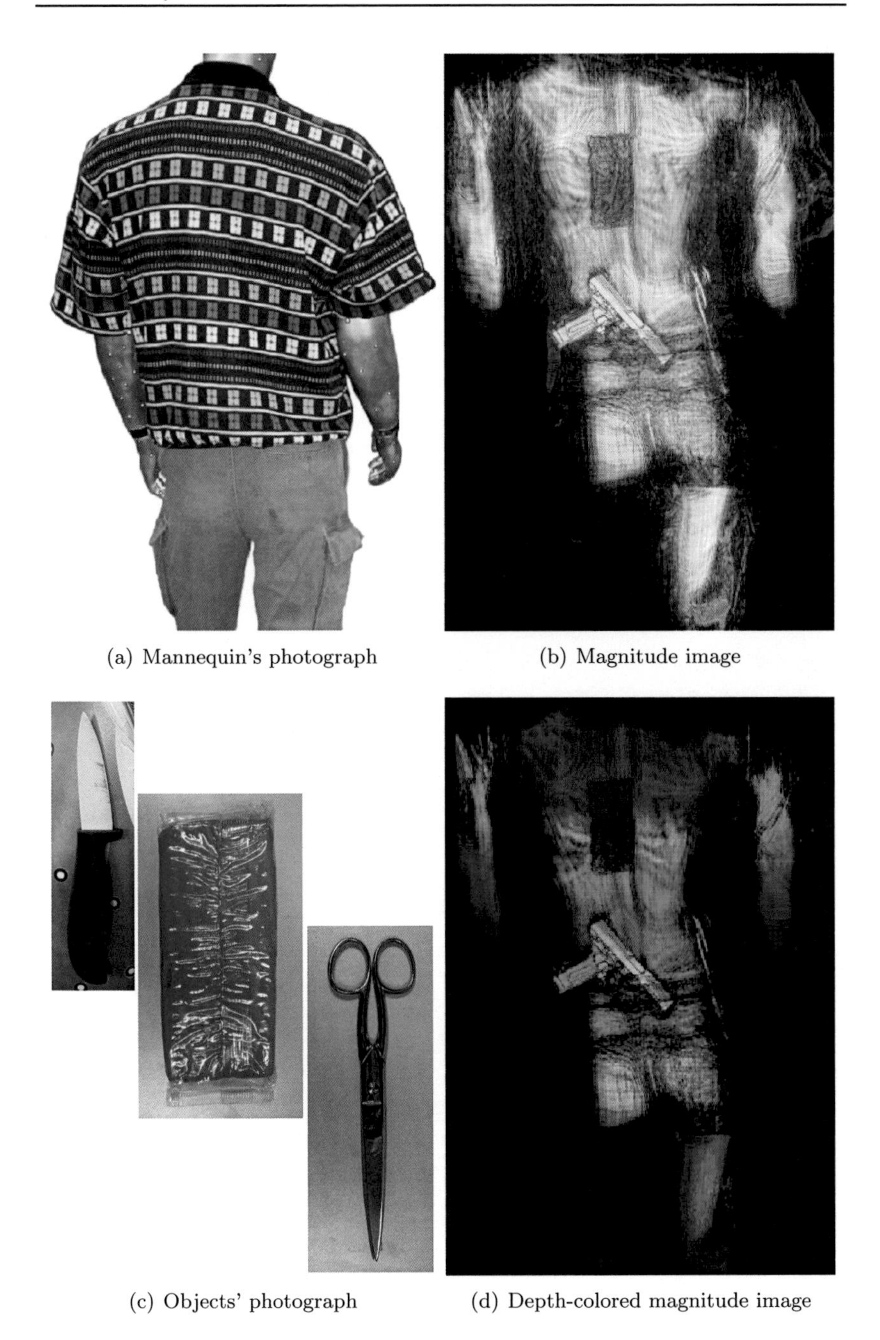

(a) Mannequin's photograph

(b) Magnitude image

(c) Objects' photograph

(d) Depth-colored magnitude image

Figure 2.29: Image of a dressed mannequin concealing various objects.

(a) Photograph

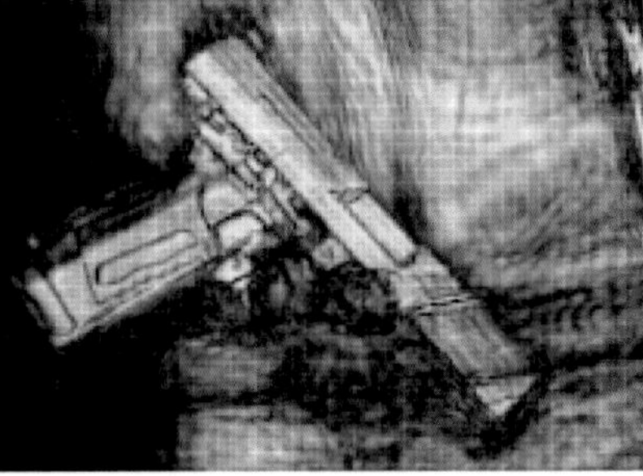

(b) Magnitude image

Figure 2.30: Detailed view of the image shown in fig. 2.29(b). The pistol is concealed behind a thick pullover and is attached beneath a leather belt.

## 2.10 Conclusion

The fundamental relations governing the operation of planar multistatic arrays have been addressed thoroughly in this chapter. Emphasis on close range operation has been made, and comparison with classical far-field formulations has been conducted where relevant. The formulation of the reconstruction process in the space domain representation as well as in the spatial frequency domain one made it possible to explore the behavior of the arrays from various aspects. Multistatic imaging delivers higher image quality than monostatic one due to the significant enhancement in signal allocation inside the K-space. Array synthesis based on effective aperture principle has been examined and adapted for multistatic arrays. Two test arrays have been introduced to operate in the millimeter-wave range. Both deliver good image quality, which validates the synthesis method.

Many practical considerations for a successful realization of multistatic imaging arrays have been discussed, and their advantages against monostatic ones have been highlighted. Multistatic arrays proved being powerful in building large arrays due to the associated reduction factor in the total number of elements, e.g., factor of hundreds. Successful design of one-square-meter array has been demonstrated along with various experimental validations for image properties and quality. In addition, scalable array geometry through clustering of antennas has been introduced in order to allow flexible array design.

Multistatic array imaging thus opens a huge opportunity for microwave imaging applications, especially the ones targeting fully electronic operation with high image quality.

CHAPTER 3

# Illumination in Reflection Imaging

---

The active reflection imaging is greatly influenced by the illumination conditions made by the geometrical arrangement of the imaging array and the imaged object. Objects with electrically smooth surfaces cause strong specular reflections, for which the incident signals are reflected within a limited angular range. Such reflections restrict the visibility of the object depending on the angle of view [79, 80]. The final quality of the illumination in the reconstructed image is influenced by both the positioning of the transmitters illuminating the imaged object as well as the positioning of the receivers collecting the reflected signals. In sparse multistatic arrays, the transmitters and the receivers are carefully positioned within the array aperture in order to complement each other, leaving less room for specific choices made for optimizing the illumination. Therefore, it is essential to understand the factors governing the illumination quality and the methods to enhance it. This chapter discusses the key factors influencing the illumination quality delivered from multistatic array imaging. Quantitative analysis of the variations in the image brightnesses due to the specular type of reflections is introduced. Simulation method suitable for the illumination problem is developed, and is afterwards applied to illustrate the case of imaging humans in the millimeter-wave range.

## 3.1 Theoretical background

The reconstruction of an image is based on the assumptions and simplifications discussed in chapter 2. This led to the application of Born approximation, with its range of validity discussed in detail in sec. 2.4. As the reconstruction method is perfectly adjusted for imaging weak diffuse reflective scatterers, it therefore fails to image probably the specular reflective surfaces. In these considerations, the antennas are assumed to fully illuminate the imaged object and thus they do not contribute to any restrictions on the illumination quality of the image. If this is not the case, then additional reductions in the illumination quality would appear.

In many applications, the imaged object consists of a combination of both smooth surfaces and diffuse reflective parts. The parts causing diffuse reflections are easier to reconstruct, as they are visible over a wide angle range. On the contrary, the surfaces causing specular reflections would be visible once they are positioned to produce a phase profile matching the phase used in the reconstruction process. The phase profile would fit exactly for the points where a specular reflection takes place, namely where the law of reflection applies. Apart from these positions, the phase deviates gradually, as detailed later in sec. 3.4.2. Consequently, the smooth surfaces of the imaged object appear bright in the reconstructed image to some spatial extend, and then suddenly turn dark.

## 3.2 Illumination boundaries

For a general planar multistatic array with a discrete distribution of transmitters $a_t$ and receivers $a_r$ on an orthogonal $[n, m]$-grid, the effective aperture $a_e$ from eq. 2.94 can be re-expressed as

$$a_e [n, m] = a_t * a_r = \sum_{\forall v} \sum_{\forall w} a_t [v, w] \cdot a_r [n - v, m - w] . \qquad (3.1)$$

For redundancy-free multistatic arrays, $a_e$ is flat; otherwise a proper weighting of the measurements of each Tx-Rx pair is needed to restore the flatness of the effective aperture. Flat effective aperture is essential for homogeneous illumination in the focused image, as addressed in sec. 3.3.

While imaging a smooth surface, specular reflections occur where the angles of incidence and reflection are equal. This is illustrated in fig. 3.1 for the simple case of a single transmitter and receiver placed on a line parallel to a large metal plate. For the transmitter at position $l_t$ and the receiver at position $l_r$, the specular point is located at $(l_t + l_r)/2$. From eq. 3.1, the effective position is located, however, at $(l_t + l_r)$. For the general case of a metal plate placed parallel to a planar multistatic array with an effective aperture $a_e [n, m]$, the illuminated region is given by the area occupied by $a_e [2n, 2m]$. It is thus convenient to define the array illumination effective aperture $a_e^{ill}$ by

$$a_e^{ill} [n, m] = a_e [2n, 2m] , \qquad (3.2)$$

which is physically interpreted as the illuminated portion in the resultant image of an infinitely large metal plate placed parallel to the array physical aperture. An illustration for the physical, effective, and illumination effective apertures is shown in fig. 3.2 for an example multistatic array.

The illumination boundary is geometrically related to the illumination effective aperture mainly through the surface topology of the scanned object and its position relative to the array. According to their geometrical

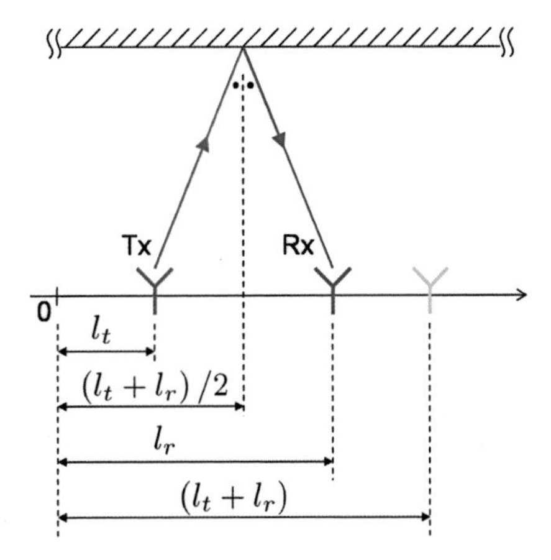


Figure 3.1: Illustration of the geometrical relation between the position of the specular point and the effective aperture position for a simple case.

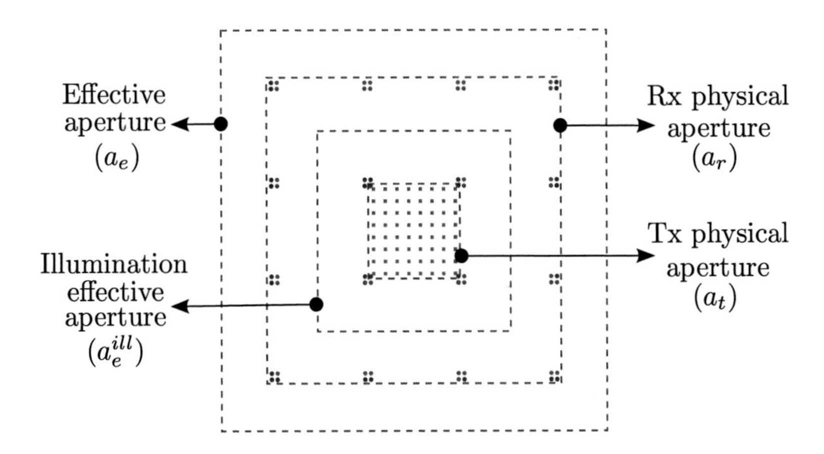


Figure 3.2: Illustration of the various aperture definitions used with planar multistatic arrays. Red and blue dots represent Tx and Rx antenna positions, respectively. The dotted lines are the boundaries of the annotated aperture definitions.

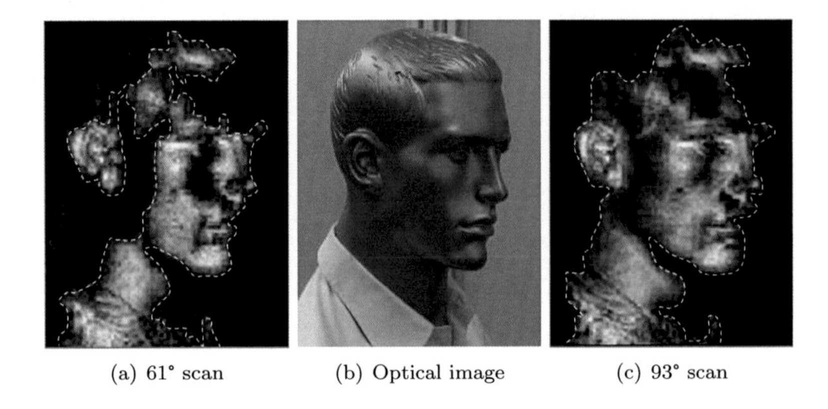

(a) 61° scan (b) Optical image (c) 93° scan

Figure 3.3: Scan results from 75 to 82 GHz of a metalized mannequin head with two different total scan angles in the horizontal plane. Dotted yellow lines illustrate the illumination boundaries [81].

arrangement, some surface portions will appear bright when they face the illumination effective aperture, and the rest keeps dark. The contours of these bright surface portions are thus defined to be the illumination boundaries. In order to visualize the influence of the illumination effective aperture on the extensions of the illumination boundaries and accordingly on the image quality of complex smooth surfaces, scan results from 75 to 82 GHz using monostatic configuration with two different aperture widths are presented. The first in fig. 3.3(a) shows the imaging result of a metalized mannequin head scanned with a total scan angle of 61° in the horizontal and vertical planes measured from the head center. And the second in fig. 3.3(c) shows the result while 93° was used in the horizontal plane instead. While increasing the total scan angle, more specular reflections from the steep surfaces coincide in the array aperture, and therefore become visible in the resultant image.

For monostatic arrays, the physical aperture and the illumination effective aperture are identical. Hence, a multistatic array and a monostatic one can yield the same illumination boundary in the resultant image if and only if the Tx and the Rx physical aperture sizes in the multistatic array are equal to the monostatic array aperture. Under such condition, the effective aperture of the multistatic array will be twice larger than its physical one in its two principle directions, and consequently its illumination effective aperture will be equal to its physical aperture and thus maximized. This does not constrain the choice of either the Tx or the Rx physical aperture distributions, but only their overall sizes.

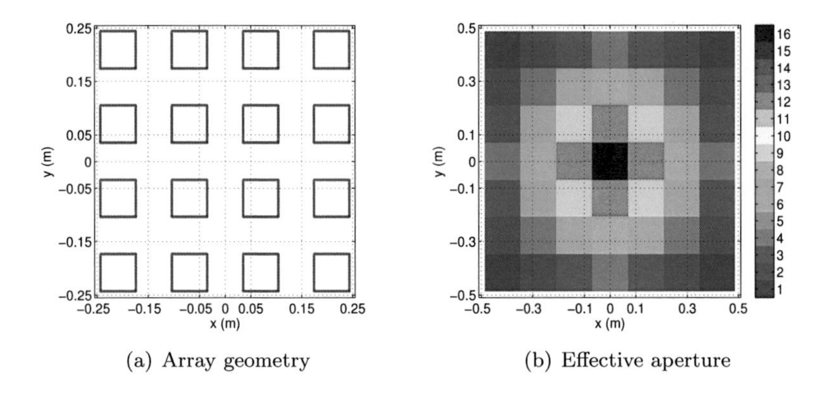


(a) Array geometry

(b) Effective aperture

Figure 3.4: Illustration of the effective aperture for an example array.

# 3.3 Illumination equalization

The effective aperture of planar multistatic arrays given by eq. 3.1 is generally not flat. The illumination effective aperture, follows from eq. 3.2, is similarly structured and hence can include abrupt changes. The smoothing of the effective aperture is thus necessary to avoid abrupt illumination changes as well as to reduce the sudden end of the aperture at its edges. Otherwise, this would lead to inhomogeneous illumination within the illumination boundaries and would degrade image quality and disturb its interpretation. Figures 3.4 and 3.5 show two examples for the effective apertures. It is therefore important to flatten the illumination effective aperture prior to image reconstruction. As access to the measurement at each Tx-Rx combination is feasible, proper weighting factors can flatten the aperture and hence the brightness levels within the reconstructed image are adjusted. It is remarkable that this adjustment cannot be performed on the resultant image from non-flat aperture, and must be applied to the reflection data before the reconstruction process takes place.

For a discrete aperture of transmitters and receivers placed on a grid where the Tx's are placed at $[n_t, m_t]$ positions and the Rx's at $[n_r, m_r]$ ones, the measurement data $s$ of eq. 2.49 can be rewritten for the discrete grid as

$$s(x_t, y_t, x_r, y_r, k) \Rightarrow s[n_t, m_t, n_r, m_r, k]\,. \tag{3.3}$$

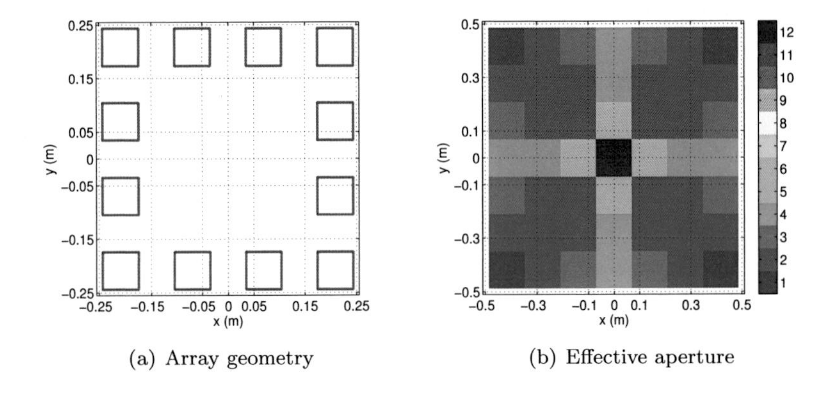


(a) Array geometry (b) Effective aperture

Figure 3.5: Illustration of the effective aperture for an example array.

The flattening of the aperture is thus described by normalizing the measurement value at the respective Tx and Rx to the corresponding value of the effective aperture. This can be further understood as an equalization process for the number of hits made by all combinations of Tx-Rx contributing to the same position on the illumination effective aperture, and hence would address same surface portions afterwards in the reconstructed image. Mathematically, this process is described by

$$s_{\mathrm{eq}}[n_t, m_t, n_r, m_r, k] = s[n_t, m_t, n_r, m_r, k] \cdot \left( \frac{w[n_t + n_r, m_t + m_r]}{a_e[n_t + n_r, m_t + m_r]} \right) \tag{3.4}$$

In eq. 3.4, the window function $w$ is also added in order to smooth the effective aperture, which can be selected depending on the target application of the imaging array. Selecting it equal to unity, the illumination effective aperture will be flattened to a rectangular shape. Such an abrupt change at the aperture edges can also degrade image quality, and hence a further smoothing might be required. Fig. 3.6 illustrates one-dimensional representation of a few choices for the window functions calculated according to [82] and [83]. Among the various types of these functions, Kaiser window with a low shape factor is favorable because it partially conserves the aperture at its edges. The exact choice of the window function $w$ will, of course, influence the lateral focusing quality of the array. This includes the side-lobe level as well as the resolution of the image, as illustrated in fig. 3.7.

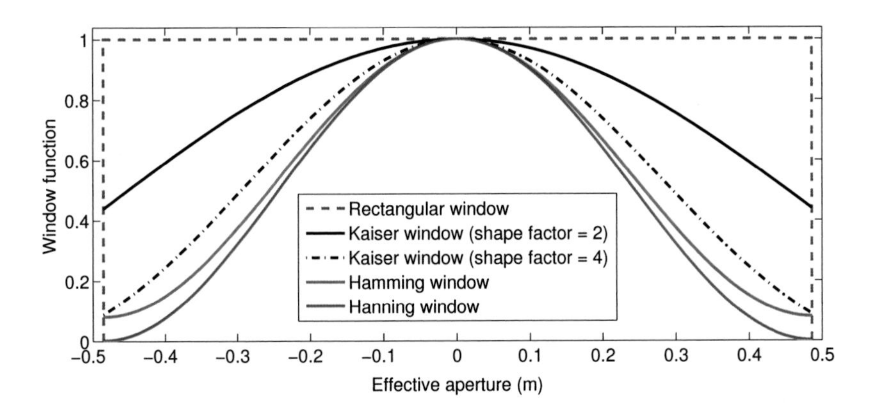


Figure 3.6: Various types of example window functions which can be used to smooth the effective aperture. One-dimensional functions are shown.

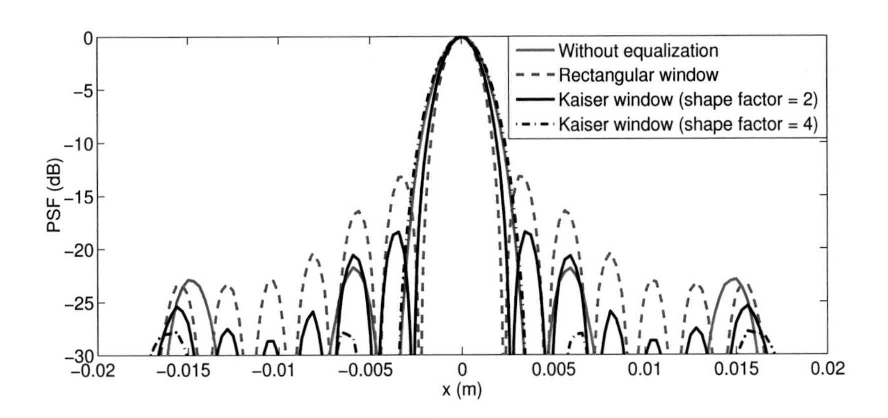


Figure 3.7: Point spread functions of the array in fig. 3.4 using different window functions seen in fig. 3.6. Simulations for 50 cm distance and a frequency range of 72 to 80 GHz are presented. Their influence on the levels of the side-lobes in lateral focusing is significant, however the change in lateral resolution is negligible.

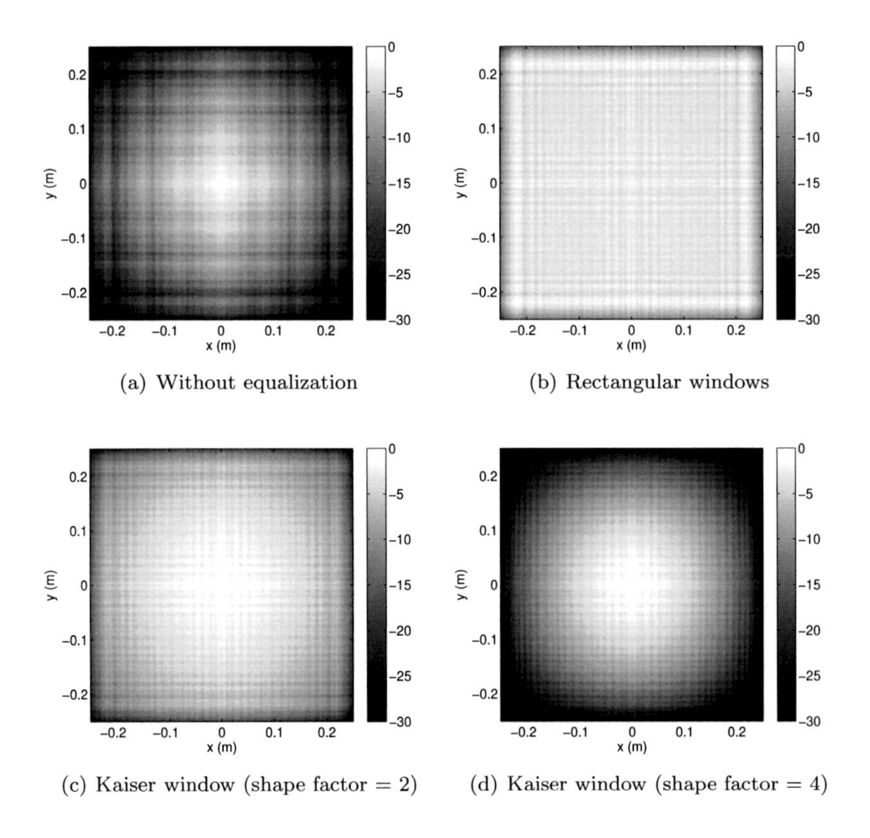


(a) Without equalization

(b) Rectangular windows

(c) Kaiser window (shape factor = 2)

(d) Kaiser window (shape factor = 4)

Figure 3.8: Image of a metal plate on a 50 cm distance to the array in fig. 3.4, simulated using different window functions. The frequency range is 72 to 80 GHz. The images are in a normalized dB scale.

For illustration, simulation results of a metal plate placed on a 50 cm distance to the array in fig. 3.4 were performed with various types of $w$ functions. In fig. 3.8(a), the image result without the flattening of eq. 3.4 is shown. The high structured illumination is clearly visible and reflects the actual illumination effective aperture of the array. By applying the equalization process, however without the window function $w$, the image is much enhanced in quality as shown in fig. 3.8(b). Further smoothing using Kaiser windows yields the results in fig. 3.8(c) and 3.8(d). Depending on application, an appropriate window function should be selected to optimize the visibility of the imaged object.

## 3.4 Processing gain

The reconstruction algorithm makes use of the coherent summation of the reflected signals out of each voxel of the imaged object. This coherent summation is achieved by the focusing terms in eq. 2.48, which is based on the reversing of the expected phase profile. The processing gain (PG) is thus defined as the magnitude of this coherent summation relative to a single measurement $s$. This is expressed as

$$\mathrm{PG} = \left| \frac{\widehat{O}(x,y,z)}{s} \right| , \tag{3.5}$$

or equivalently in dB as

$$\mathrm{PG}_{\mathrm{dB}} = 20 \log_{10} \left( \left| \frac{\widehat{O}(x,y,z)}{s} \right| \right) . \tag{3.6}$$

The understanding of the PG produced by the imaging system is necessary for proper interpretation of the different brightness levels produced in the resultant image, which has been an important issue since the early surveillance imaging radar systems [79]. The PG is also a key factor in the imaging system because it plays a major role in the enhancement of the signal-to-noise ratio after focusing.

On one hand, the focusing terms fit perfectly the diffuse reflections. On the other hand, the specular reflections will produce a phase profile which would fit partially with the focusing terms, and therefore less PG is achievable. In the following, the PG produced by a simple planar multistatic array is analyzed. For the sake of simplicity, the signal variations due to illumination equalization of eq. 3.4, free space attenuation, and antenna radiation patterns are not considered. The results are presented for a single frequency of operation, however can be easily extended to cover the case of multiple frequencies by linearly scaling the PG to the number of frequencies used.

### 3.4.1 Diffuse reflections

Rough surfaces, edges, and small objects relative to the applied wavelength cause the incident signal to scatter back to all receivers almost isotropically and hence the PG is simply deduced from eq. 3.5 as

$$\mathrm{PG} = N_t \cdot N_r , \tag{3.7}$$

where $N_t$ and $N_r$ are the total number of the used transmitters and receivers, respectively.

In practice, shadows can hinder achieving this gain; and object sizes close to or greater than the wavelength do not produce complete isotropic scattering. Therefore, the value given by eq. 3.7 is the theoretical maximum for the achievable PG.

### 3.4.2 Specular reflections

Electrically smooth surfaces have extensions larger than $\lambda$, however with a surface roughness $\ll \lambda$. They thus produce specular reflections and cause a phase mismatch to the focusing terms as shown next. In order to find the PG under specular reflection condition, the case of an infinitely large metal plate placed parallel to the array aperture on a distance $R$ is considered. A simple square planar multistatic array consisting of a 2D equidistant grid of receivers and one central transmitter is thus used to image the plate. The array is centered in the $xy$-plane, i.e., the Tx is placed at $(0, 0, -R)$. Using the image theory [84], the measurement $s$ collected by a receiver at a position $(x_r, y_r, -R)$ is given by

$$s = |s_o| \cdot e^{-jk\sqrt{x_r{}^2 + y_r{}^2 + 4R^2}}, \tag{3.8}$$

where $s_o$ denotes the magnitude value including the free space attenuation and the gain of the used antennas.

Then, the focusing terms of eq. 2.48 for the target voxel $(x, y, 0)$ on the metal plate reduces to

$$\underbrace{e^{+jk\sqrt{x^2+y^2+R^2}}}_{\text{Tx focusing}} \cdot \underbrace{e^{+jk\sqrt{(x_r-x)^2+(y_r-y)^2+R^2}}}_{\text{Rx focusing}} . \tag{3.9}$$

Considering a target voxel at $(0, 0, 0)$, and by normalizing the phase terms to the wavelength, the PG after the reconstruction is expressed as

$$\text{PG} = \left| \sum_{\forall \hat{x}_r} \sum_{\forall \hat{y}_r} \underbrace{e^{-2\pi j\sqrt{\hat{x}_r^2+\hat{y}_r^2+4\hat{R}^2}}}_{\text{slow}} \cdot \underbrace{e^{+2\pi j\sqrt{\hat{x}_r^2+\hat{y}_r^2+\hat{R}^2}}}_{\text{fast}} \right| . \tag{3.10}$$

The $\hat{\cdot}$ symbol indicates a value normalized to the wavelength. By observing the two phase terms in the PG expression, it is evident that one is a fast varying term, and the other is a slow varying term with respect to the lateral position of the receiver. This fact will be used to further simplify the expression in the following analysis. The 2D phase profile of the individual signals described by eq. 3.10 per receiver position is illustrated in fig. 3.9(a) for an example case of $\hat{R}$ equal to 128. Due to the phase mismatch between the measurements and the focusing term, only the inner zone of the phase profile in fig. 3.9(a) contributes to the PG constructively, while the rest produces an oscillating phase profile which adds destructively.

In order to illustrate the behavior of the reconstruction using a numerical example, the spacing of the receiver grid is chosen to follow aliasing-free spatial sampling, namely $\lambda/2$. The PG is hence calculated for square apertures of side lengths up to $150\lambda$ by fixing $\hat{R} = 128$. Fig. 3.9(b) shows

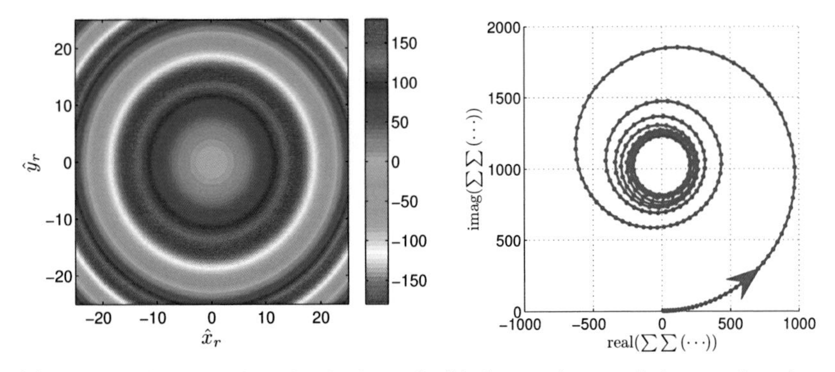


(a) Phases in degree of the individual signals summed for the PG, indicating constructive summation in the aperture middle region.

(b) Summation result in complex plot. Arrow indicates the curve growth while increasing the aperture area.

Figure 3.9: Illustration of the phase and the complex summation results of the signals contributing in the PG after specular reflections on a metal plate placed on a $128\lambda$ distance from the array.

the result of the summation in eq. 3.10 in a complex plot while increasing the side length of the aperture as with the shown arrow direction. The PG is therefore the absolute value of this spiral trace[§], which asymptotically reaches a final point at approximately $(0, 1000)$ or equivalently 60 dB. This asymptotic value is therefore defined as the average PG. Having an imaging array with dimensions relatively larger than the size of the inner zone yields the PG to reach its asymptotic value. Fig. 3.10 shows the PG for the case of diffuse scattering and specular scattering. For large arrays, the PG varies significantly for diffuse and specular reflective objects.

The average PG is dominated by the signals within the inner zone region of the phase profile. As a matter of approximation, the effective extension of the inner zone is considered to be where the phase variation is $\leq \pi/2$. This can be graphically interpreted as the condition where the added vectors start to build the first overshoot in the PG, after which the first destructive addition takes place. By neglecting the slow varying part of eq. 3.10, the condition is formulated as

$$2\pi\sqrt{(\hat{d}/2)^2 + \hat{R}^2} - 2\pi\hat{R} = \pi/2\,, \tag{3.11}$$

[§]The reader might notice the geometrical similarity between this spiral curve and the Cornu spiral known for the solution of Fresnel integrals. This spiral curve is, however, not a Cornu spiral due to the discrete nature of the summation, the different scaling in the phase factor, and the usage of square aperture. Solutions of Fresnel integrals are often used in tabular forms prepared mainly for optical systems [85]. They also often include several mathematical simplifications.

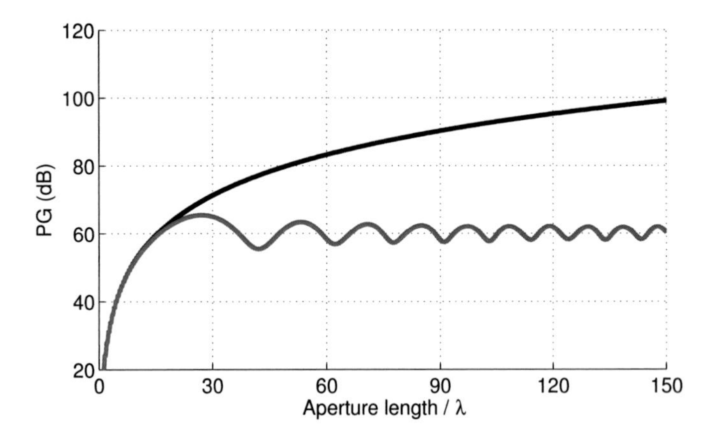


Figure 3.10: Processing gain achieved with a square aperture for diffuse reflection (upper black line) and specular reflection (lower blue line) on a metal plate placed on a $128\lambda$ distance from the array.

where $\hat{d}$ is the normalized diameter of the effective inner zone. This yields the description of $\hat{d}$ to be

$$\hat{d} = \sqrt{2\hat{R} + 1/4} \,. \tag{3.12}$$

The constraint in eq. 3.12 is now applied to eq. 3.10 in order to find the average PG for several $\hat{R}$ values with a moderate numerical effort. The numerical results of these calculations are shown in fig. 3.11 using different antenna spacings. On one hand, increasing $\hat{R}$ will increase the PG because the phase front migrates towards a plane wave and hence the inner zone diameter extends. On the other hand, the illumination boundaries on curved objects will shrink, causing an overall reduced illumination result.

For a general position $(x, y, 0)$, the phase profile will shift spatially and the PG would not change much. The center of the inner zone is the point fulfilling the law of reflection. Therefore, the reconstruction results in homogeneous brightnesses up to the illumination boundaries, after which the inner zone migrates outside the boundaries of the physical aperture and leaves the oscillating phase profile instead. Consequently, the PG vanishes and dark zones in the resultant image appear.

For the more general case of smooth non-planar surfaces, the center of the inner zone follows the law of reflection, and the diameter of the effective inner zone will be modified by the object local radius of curvature. Surfaces of the imaged objects used in practice range from planar to convex. The latter produce larger zones in the phase profiles than the former,

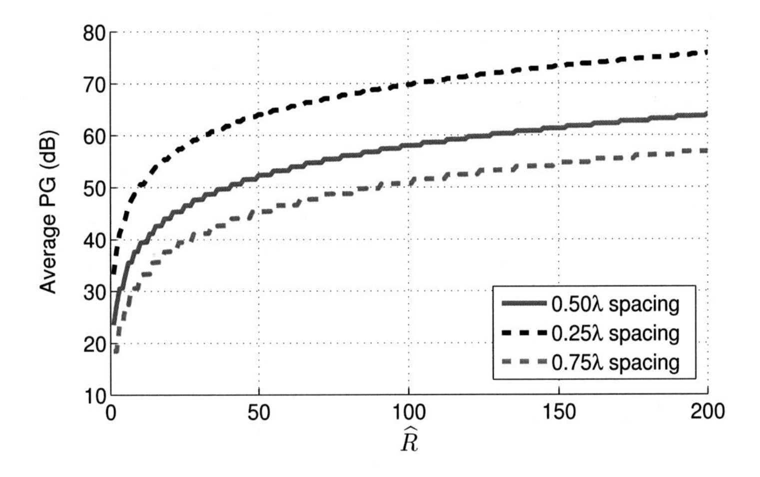


Figure 3.11: Values of the average PG dependent on the distance $\hat{R}$ for square multistatic arrays of different antenna spacings.

and therefore receive higher PG. Thus, the two lines shown in fig. 3.10 introduce the theoretical minimum and maximum for the available PG, at the considered $\hat{R}$ distance.

### 3.4.3 Experimental verification

A verification measurement is set up in order to examine the PG factors. Two test objects producing diffuse and specular reflections, fig. 3.12(a), are prepared for comparison. The first one is a 30 mm long aluminum stick of 4 mm diameter ending with a round tip and is prepared with an absorbing material on its background. The monostatic RCS of the metal stick is −49 dBsm at 80 GHz known from a commercial field simulator. The second one is a 500 mm times 500 mm metal plate placed behind the metal stick. Both are mounted in front of a mechanical scanning array on 480 mm distance, which is 128$\lambda$ at 80 GHz. Due to the free space attenuation and the different RCS's, $s_o$ is 50 dB higher for the metal plate than for the metal stick. Measurements from 75 to 85 GHz with the multistatic array introduced in sec. 3.4.2 are collected. The array physical aperture is 480 mm times 480 mm. The 10 GHz bandwidth is used to enhance range focusing. The measured data is hence applied to eq. 2.49 for focusing. A slice cut of the reconstructed 3D image is shown in fig. 3.12(b), which illustrates the brightness values of both objects.

Using the multistatic array and the selected scan geometry, excess average PG of 36 dB for the metal stick relative to the metal plate is

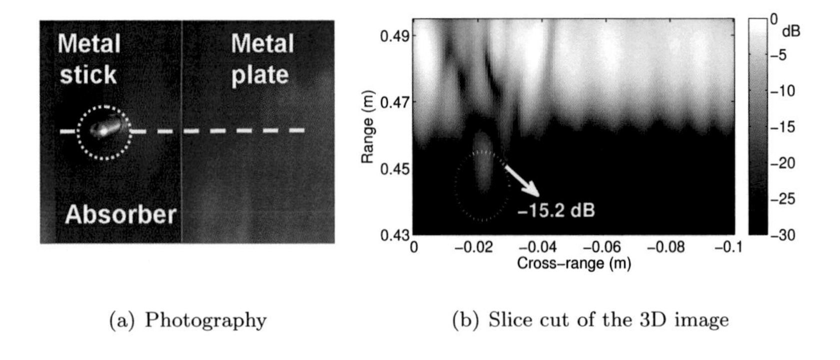


(a) Photography

(b) Slice cut of the 3D image

Figure 3.12: Imaging result using 75 to 85 GHz scan at 48 cm distance showing the brightness variations due to diffuse and specular reflections. The dotted line shows the plane of the presented slice cut.

expected, according to the results in fig. 3.10. This results in an expected brightness difference of $(50-36) = 14$ dB more for the metal plate relative to the metal stick. The imaging result shows brightness difference of 15.2 dB, thus a good agreement with the theoretical calculations is met. This small deviation of 1.2 dB is due to the unconsidered variations in the free space attenuation and the radiation patterns of the used antennas.

## 3.5 Simulation method

It is very helpful, while designing imaging arrays, to observe and study the reachable illumination boundaries delivered by the array geometry. Theoretically, the problem can be modeled using electromagnetic field solvers to simulate for object reflections, followed by the image reconstruction step. In practice, however, this process is hardly realizable because of the huge electrical problem caused by the imaged object, i.e., typically larger than $(100\lambda)^3$ volume is to be simulated. Additionally, field solvers must calculate the reflections for each transmitter and frequency, thus repeating the task many times, making the problem impractical to solve, especially for large arrays. A simplified, yet accurate, simulation method is therefore very demanded to ease array design and evaluation.

Typically, it suffices to differentiate the bright and dark regions expected in the image according to the geometry of the imaged object. For objects of highly reflective surfaces, the simulator must calculate the possibility to reconstruct each portion of that surface by the proposed geometry. Thus, knowing the 3D model of the object surface, the array geometry, their relative positioning, and the frequency of operation is significant for an accurate simulation. Due to the nature of the reconstruction process and the associated phenomena discussed in sec. 3.4, a simple ray-tracing technique, as used in the optical domain, would not work correctly. Ray-tracers search for an exact geometrical solution for the reflected rays, which should hit the receiver precisely. This is applicable at the limiting case of infinitesimally small wavelengths, e.g., at the optical frequencies. At microwave and millimeter-wave frequencies, the concept of rays must be relaxed due to the longer wavelengths used.

Following the results in sec. 3.4, the effective inner zone of the phase profile available on the receiver array has a normalized diameter $\hat{d}$ defined in eq. 3.12. The reflected rays, which reach any receiver within the region defined by $\hat{d}$, would contribute to the illumination result. The region described by $\hat{d}$ is the volume confined by a paraboloid which extends with the normalized distance $\hat{R}$. This condition is illustrated in fig. 3.13 for an example voxel $O$ where a specular reflection takes place. The incident ray $\vec{t}$ is reflected in the direction $\vec{r}$ following the law of reflection. The vectors $\vec{P_{to}}$ and $\vec{P_{or}}$ show the directions from the transmitter and to the receiver relative to $O$, respectively. The voxel $O$ is hence visible[§] in the image only and only if the vector $\vec{P_{or}}$ lies within the paraboloid confined by the diameter $\hat{d}$.

Modeling large electrically smooth objects for imaging purpose is best achieved by using 3D surface models. Such models are based on surface meshing techniques known in computer graphics [86], i.e., by using triangulation meshing method. It describes the surface by triangles of different sizes following the surface topology to the required accuracy. Each triangle is hence defined by three points and an orientation vector, either in inward or outward direction. The task of the proposed simulator is to examine each triangle to be either visible or not at the given frequency of operation according to the illumination condition in fig. 3.13.

In order to verify the proposed simulation method, a 3D model of a mannequin was generated using a commercial optical scanning system. This model is then made available to the simulator and adjusted to be positioned against the multistatic array presented in sec. 2.9. The array consists of 16 clusters covering almost an area of 1 m times 1 m and operates from 70 to 80 GHz. The mannequin was placed on one meter distance from the array surface. The surface of the mannequin was metalized in order to ensure similar reflectivity as expected from a human

[§]The geometrical relations discussed before assumed that aperture is parallel to the surface at the specular point. For the case of very tilted surfaces and very large apertures operating at close distances, this assumption is partially falsified. It covers, however, most of imaging scenarios expected in practice.

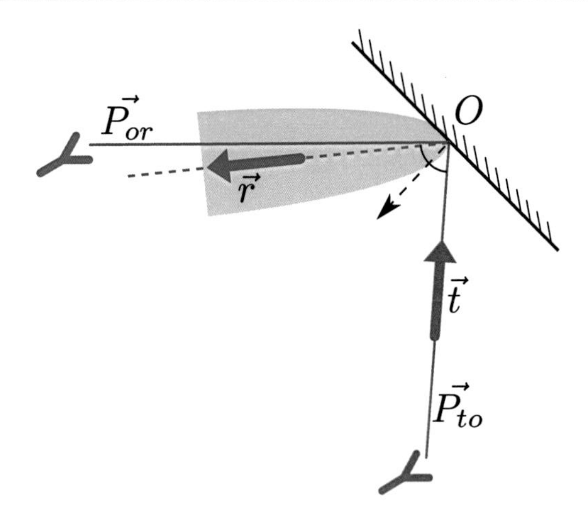


Figure 3.13: Geometrical model describing the vectors used for the illumination simulation. The voxel $O$ is visible in the image as long as the vector $\vec{P_{or}}$ lies within the paraboloid region marked in blue.

skin [87, 88]. Fig. 3.14 shows the dressed mannequin, which in addition hides a few objects beneath the clothes. The surface model as well as the normal vectors at each surface portion are illustrated for the rear view.

The mannequin is measured twice, i.e., for front and rear views. The reconstructed images after the measurements are shown in fig. 3.15(a) and fig. 3.15(c). The abrupt change in the brightness is well observed at the illumination boundaries. In fig. 3.15(b) and fig. 3.15(d), the corresponding simulation results are presented. The results are visualized using two colors, white and dark gray. The white color shows the regions satisfying the illumination condition and hence are expected to be visible in the reconstructed image, and the dark gray is for the rest of the 3D surface failed to satisfy the condition. By comparing the microwave images and the simulation results, a strong similarity is observed, which hence demonstrates successful modeling of the illumination problem and a suitable simulation method.

For the simulation cases where very large number of transmitters and receivers are involved, it could be numerically very costly to examine the illumination condition for each Tx-Rx combination. It is possible then to relax the illumination condition and replace it with direct search technique. In this, the normal vectors at each surface triangle is examined to see if it crosses the illumination effective aperture of the used array. If it does, then the triangle is most probably visible in the image, otherwise it appears dark. This method is much simpler than the one proposed above, however also less accurate.

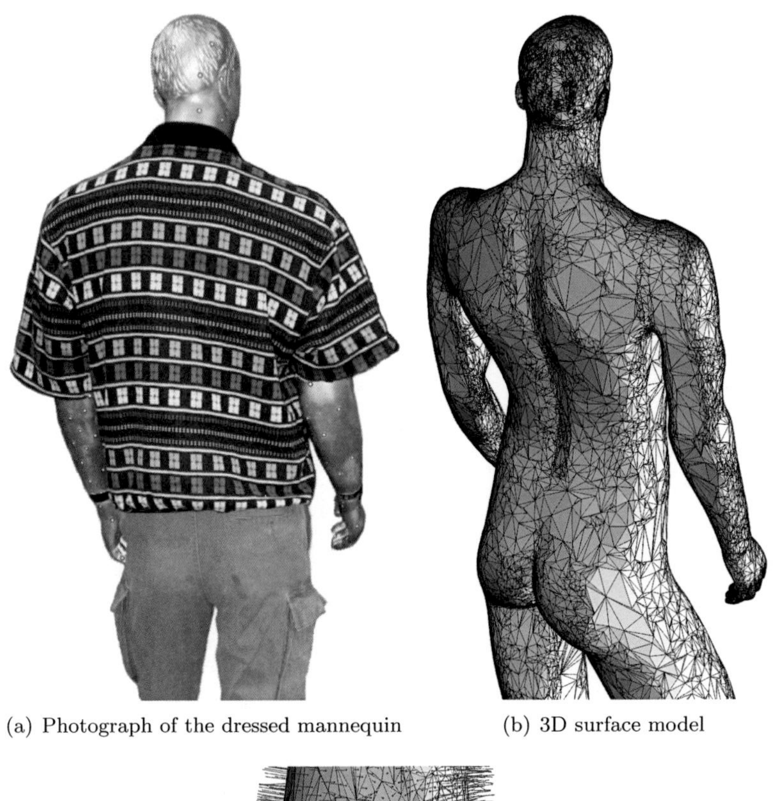

(a) Photograph of the dressed mannequin (b) 3D surface model

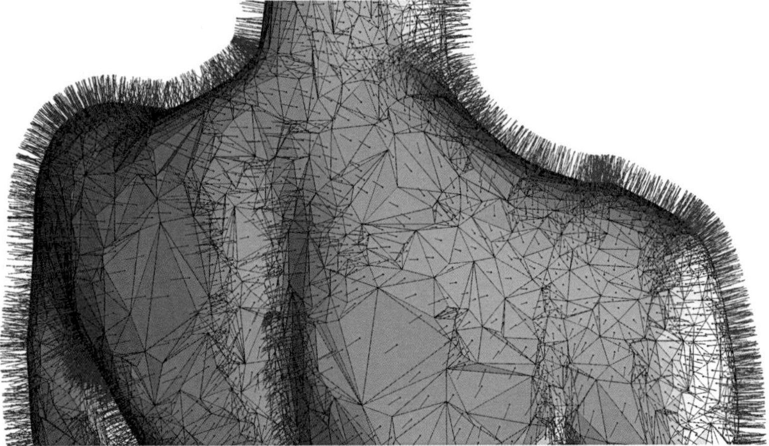

(c) Detailed view showing the surface meshing and the normal vectors

Figure 3.14: View for the test mannequin used to verify the illumination simulation method.

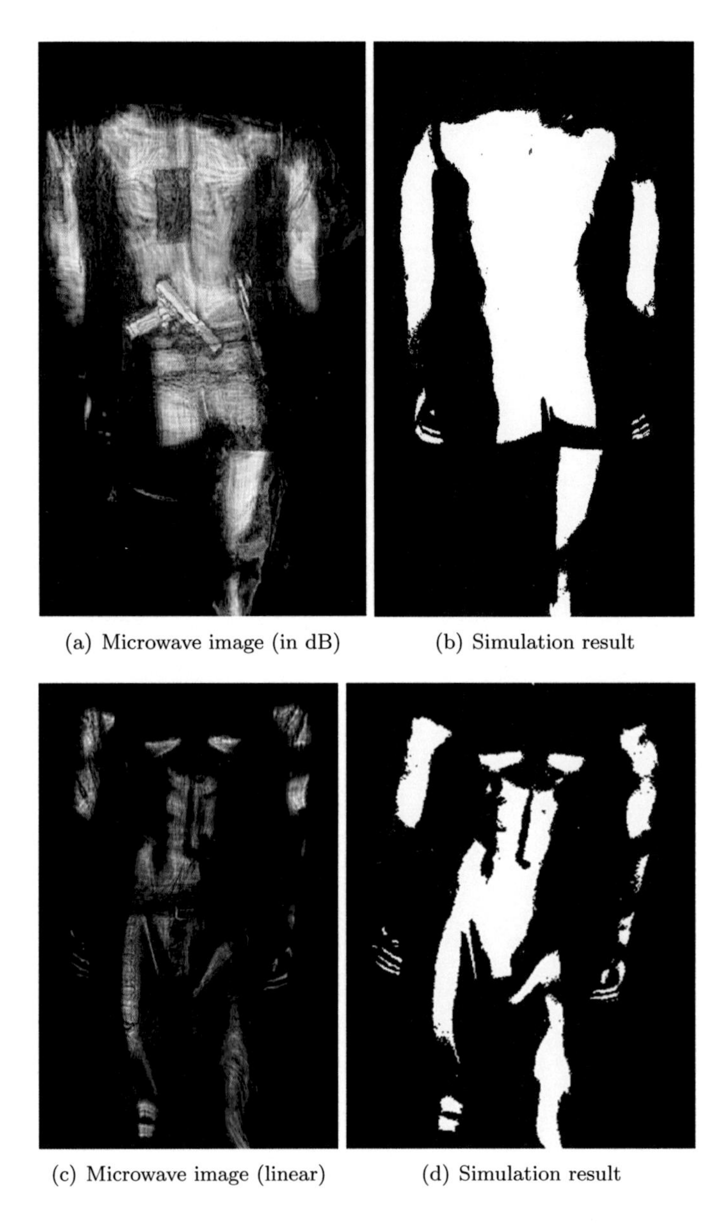

(a) Microwave image (in dB) (b) Simulation result

(c) Microwave image (linear) (d) Simulation result

Figure 3.15: Left: The microwave images measured from 70 to 80 GHz revealing some concealed objects and illustrating the illumination boundaries. Right: The corresponding simulation result of the 3D model, where the white color shows the regions satisfying the illumination condition. Simulation results clearly match the experimental ones.

## 3.6 Illumination on human body

Imaging of humans is very demanded in security and medical applications, e.g., personnel screening. The human skin is rich of water content, which makes it very reflective for millimeter-waves [87–94], causing strong specular reflections as discussed above. Therefore, active imaging of humans often suffers shadows and dark zones in reflection imaging due to the complex nature of the geometry of the body. In this section, dedicated simulations for this case are addressed.

The simulation model introduced above is used to investigate a typical scenario for imaging humans at 80 GHz. Fig. 3.16 illustrates a 3D model of a female body. The body is approximately 1.6 m tall, selected to represent an average case. This model is simulated against a planar aperture of two meters height and one meter width, which is placed approximately 80 cm away from the center of the 3D model. The array is of the same topology as in sec. 2.9, however the result is valid for any other array geometry of the same effective illumination aperture.

The 3D surface model is rotated and positioned in front of the imaging array at various angles. Two simulation results are shown in fig. 3.17 and fig. 3.18 for the front and rear views, respectively. Fig. 3.19 and fig. 3.20 show the results for the right and left views as well. In some cases, slant views can help covering other parts of the body, as illustrated in fig. 3.21 and fig. 3.22 for 45° and 60°, respectively. The results illustrate a successful coverage of most of the skin surface. Some regions appear permanently dark due to the limited aperture extension. For instance, the shoulders and the upper part of the head are especially difficult to cover, as the normal vectors in their region point vertically. For this, the array elements, and hence the illumination effective aperture, must reach the region above the body in order to make these surface parts visible. The legs are relatively difficult to cover completely due to shadowing effects caused by side views.

It should be emphasized again that the dark regions of the body are not dark due to a lack of illumination from the transmitters, however, due to the specular reflections causing the return signals to reflect away of the available receivers. In order to overcome this limitation, the body should be imaged in a possibly close distance to the imaging array and from various perspective views. Close range imaging is an essential factor in this case, where far-field active imaging in the millimeter-wave range will completely fail to image the body probably. Depending on the application in question, the placement of the array elements and the its positioning against the imaged person have to be specifically optimized.

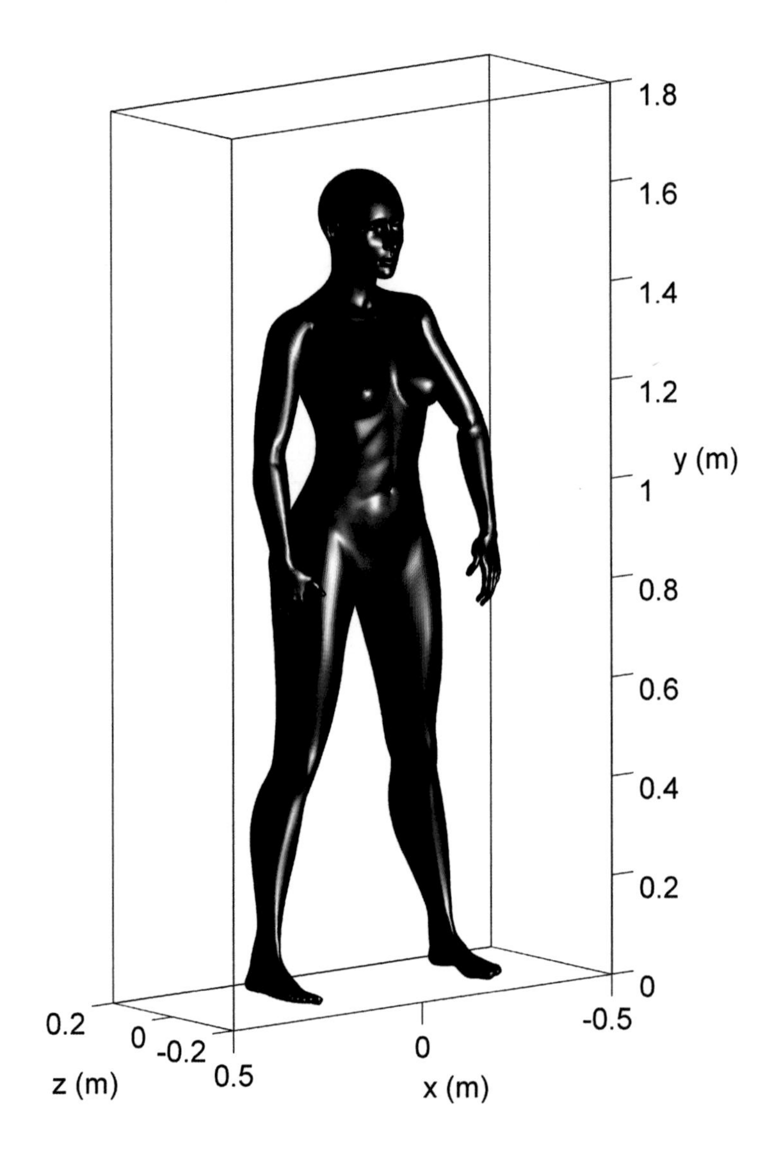


Figure 3.16: A 3D surface model of a female body used for the illumination simulations on humans.

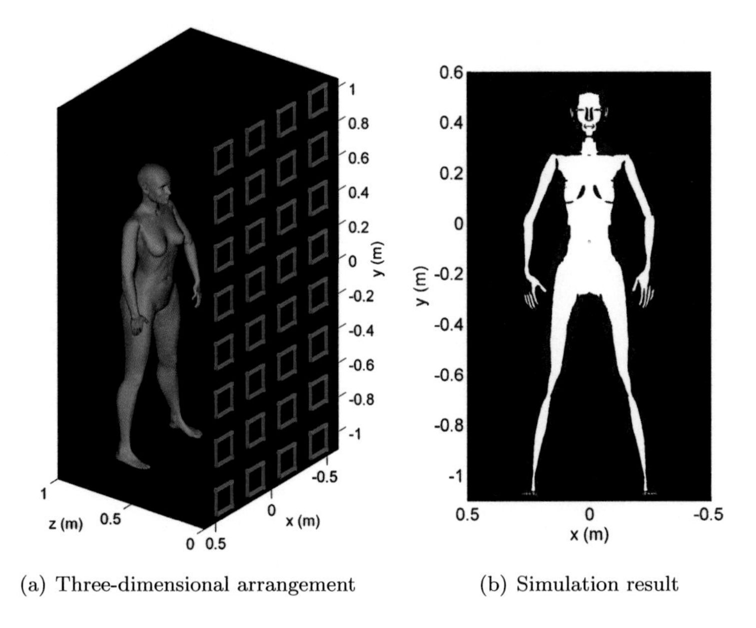


(a) Three-dimensional arrangement (b) Simulation result

Figure 3.17: Simulation result for a front view.

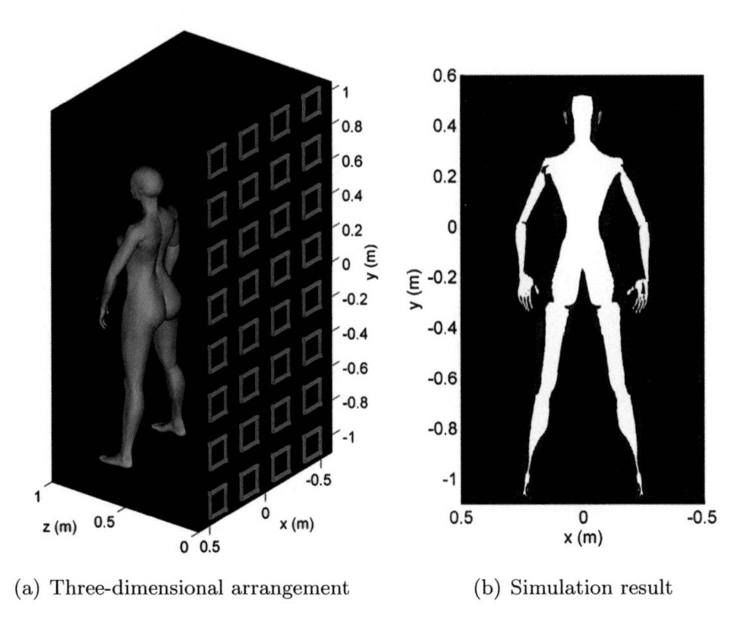


(a) Three-dimensional arrangement (b) Simulation result

Figure 3.18: Simulation result for a rear view.

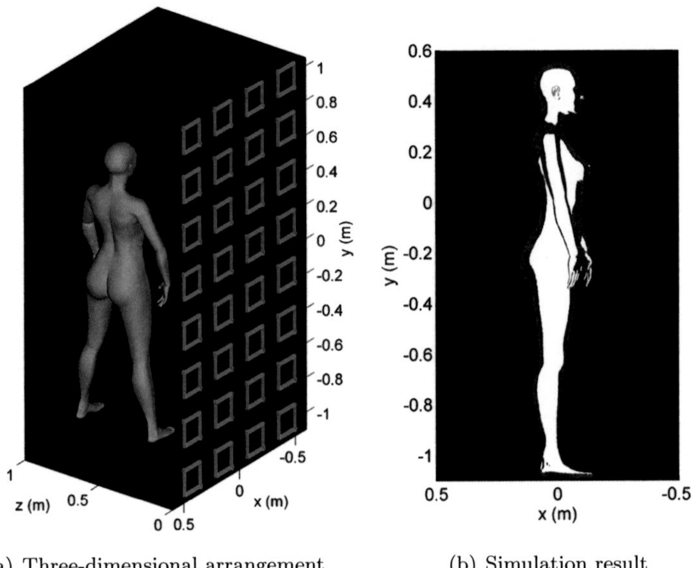


(a) Three-dimensional arrangement (b) Simulation result

Figure 3.19: Simulation result for a right view.

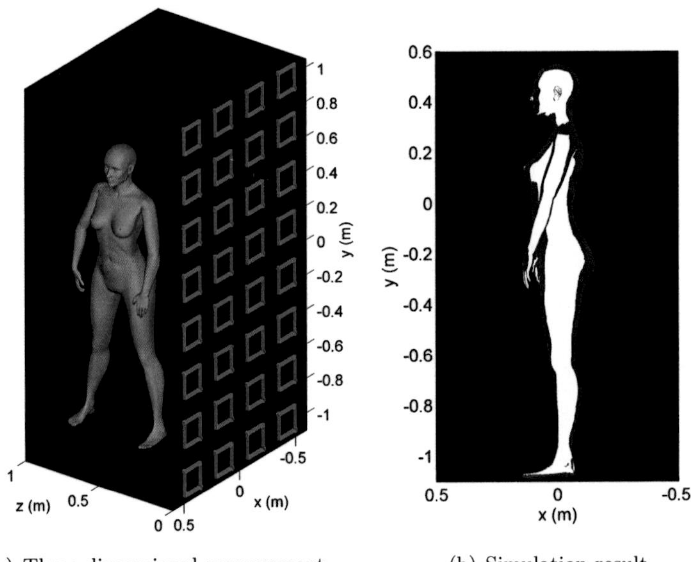


(a) Three-dimensional arrangement (b) Simulation result

Figure 3.20: Simulation result for a left view.

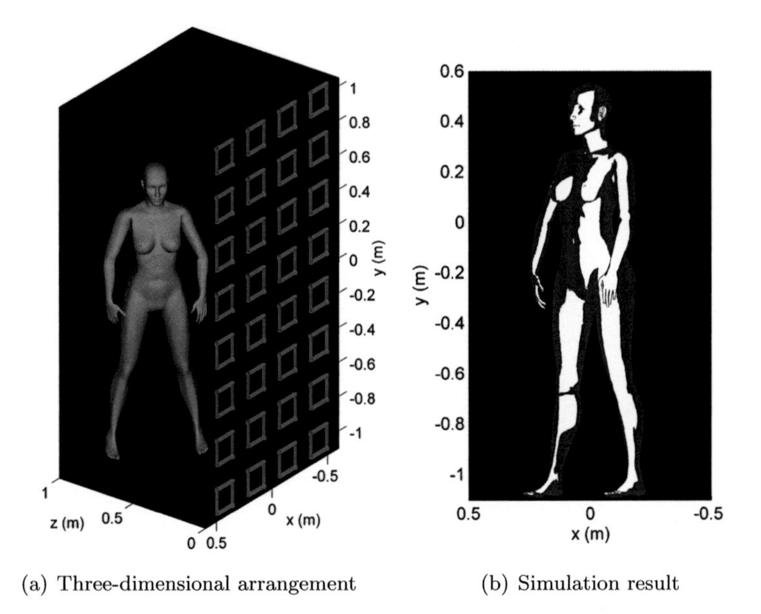


(a) Three-dimensional arrangement (b) Simulation result

Figure 3.21: Simulation result for a 45° view.

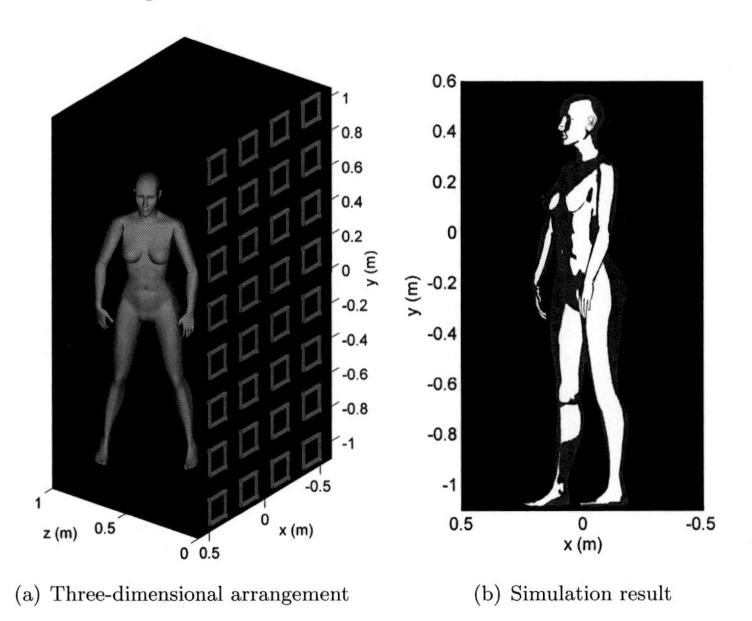


(a) Three-dimensional arrangement (b) Simulation result

Figure 3.22: Simulation result for a 60° view.

## 3.7 Conclusion

The illumination issue in active imaging using planar multistatic arrays has been analyzed thoroughly in this chapter. The associations between the illumination result in the reconstructed image and the geometrical conditions made by the array and its positioning relative to the imaged object have been detailed. New terms have been defined to describe the physical effects observed in the images, namely the illumination boundaries, the illumination effective aperture, and the processing gain. All play important roles in the understanding of the illumination issue. Illumination equalization has been introduced and is considered a specific processing step necessary for multistatic arrays. Multistatic arrays can deliver equivalent illumination coverage as the monostatic ones once the illumination equalization step is performed. The mathematical formulation for this equalization combined with an optional aperture smoothing with a window function has been introduced along with several examples. It is the capability of the digital-beamforming technique to allow for such processing steps to be realizable.

Moreover, quantitative analysis for the achievable brightness levels inside the image has been conducted while differentiating between the two fundamental scattering phenomena, namely the diffuse and specular ones. By tracing the signal paths and the corresponding reconstruction results, the limiting values for the processing gains could be calculated, which have been verified experimentally as well. Knowing these limits could assist the design of imaging systems benefiting from the multistatic operation. In addition, proper interpretation of the images becomes feasible. Based on these foundations, a dedicated simulation method has been introduced in order to help in the evaluation of electrically large problems, especially where large arrays are also deployed. Experiments verified the results to a great extend. This method has been thus utilized to predict the illumination result achievable while imaging humans, e.g., for personnel screening applications. Body zones have been accordingly recognized to be difficult to cover probably due to their geometrical nature. The value of close range imaging is to be emphasized in helping enhancing the illumination quality in the images.

In general, illumination in image can be extended by using conformal array solutions adapted to the geometry of the imaged object. Another strategy to enhance the illumination result is to combine the array with mirrors around the imaged object [95]. Incident signals can be redirected towards the object or the reflected signals towards the receivers, or both. This is surely associated with more effort in image reconstruction, however possible to achieve. Such measures have to be specifically optimized for the application in question.

CHAPTER 4

# Calibration of Multistatic Arrays

The main task of the calibration procedure is to establish a hypothetical reference surface suitable for the reconstruction algorithm, and hence for the image production. The reconstruction algorithm assumes that the signals are directly emitted from the phase centers of the transmitting antennas and directly received at the phase centers of the receiving antennas. In practice, this is not the case, and instead, the signals undergo attenuations and time delays which are antenna dependent as well as frequency dependent. The phase relations between the array elements are of particular importance for the operation of digital-beamforming (DBF) systems, and care is therefore taken to characterize the phase relations with best possible accuracy. These relations are fully deterministic in nature, however arbitrary, where they depend on the hardware components and the frequency of operation. It is the great advantage of the DBF systems to allow for such arbitrary relations; and therefore this leads to a great flexibility in the realization of the hardware components and connections without the need for phase shifters or gain control elements in the front-ends. Instead, the establishment of the reference surface can be achieved numerically. Considering planar arrays, the reference surface is optimally chosen to be the plane including the phase centers of all transmitter and receiver antennas.

In principle, the calibration standards could be either implemented behind or in front of the antennas. Calibrating behind the antennas can be realized using either identical reference channels to the actual ones, or by means of implemented switches or splitters which guide the signal to the antenna or to a reference path. Calibrating in front of the antenna means that the calibration standard is placed in free space, or in other words, in the imaged volume. Calibration methods made behind the antennas benefit from the high signal-to-noise ratio (SNR) available, however are often less accurate than the calibration methods in free space. Firstly, the antenna characteristics are not taking place in the calibrated path, and hence are not fully corrected. Secondly, the signal paths can hardly be made exact between the reference path and the actual one. On the contrary, calibration in free space does identically include the same signal paths and hardware components used later in the imaging operation. It also does not cause any extra costs in the hardware realization for integrating extra components. For these reasons, calibration in free space is considered to be the favorable method for imaging arrays. The associated reduction in the SNR can be compensated by intelligent processing methods, and additionally by allowing a longer integration time than usually used in measurements made for the imaging purpose.

In this chapter, a suitable system error model for multistatic imaging systems is first discussed. Afterwards, a system error correction method based on two standards, i.e., match and offset-short, is presented, and then followed by a discussion regarding the practical aspects of the calibration of multistatic imaging systems. Later, a detailed discussion on the numerical simulation associated with the characterization of the reflector used to realize the offset-short standard, namely a sphere, is made. Lastly, methods for the enhancement of the calibration quality to overcome possible high noise levels and mechanical positioning errors will be presented.

## 4.1 Error model

A multistatic array generally consists of several transmitter (Tx) and receiver (Rx) antennas, connected each to a channel. Similarly as for any RF channel, each channel in the multistatic array causes time delay and magnitude change to the transmitted as well as received signals. The channels are designed to behave linearly over the signal power levels of interest. Accordingly, the systematic errors caused by the channels can be corrected using linear models. This correction procedure is therefore named system error correction[§]. The linearity is assumed separately for each frequency, and hence the system error correction must be repeated for each frequency of operation. For the multistatic case, as illustrated in fig. 4.1, the systematic sources of errors occur mainly due to

- the intrinsic characteristics of each Tx and Rx channels
- the cross-coupling between the Tx and Rx channels
- the mutual coupling between the various Tx channels
- the mutual coupling between the various Rx channels

[§]The same procedure is well-known in the calibration of vector network analyzers.

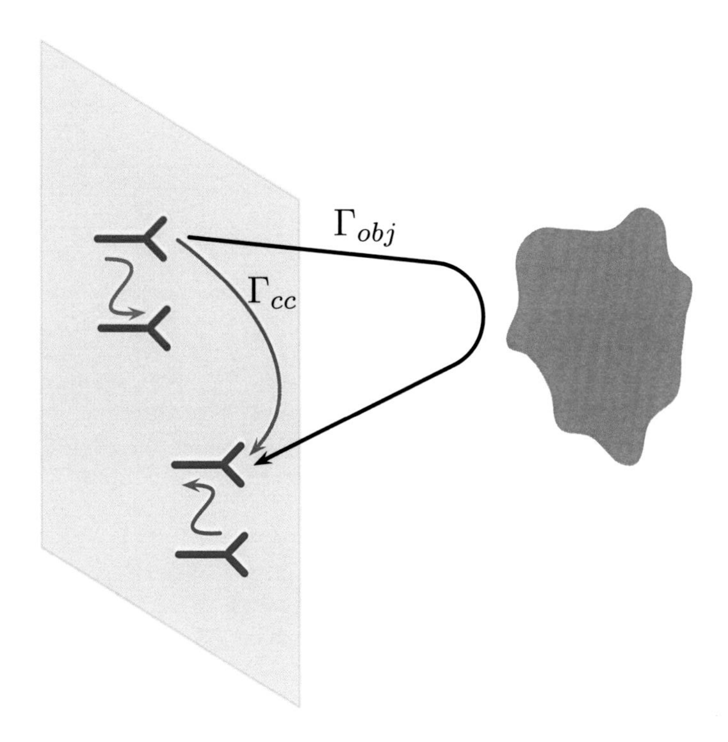


Figure 4.1: The signal flow in a multistatic array for a single measurement. Besides the demanded signal $\Gamma_{obj}$, a direct cross-coupling signal $\Gamma_{cc}$ exists, and additionally, unwanted signals caused by mutual coupling within the Tx channels (marked in red) and within the Rx channels (marked in blue) are present.

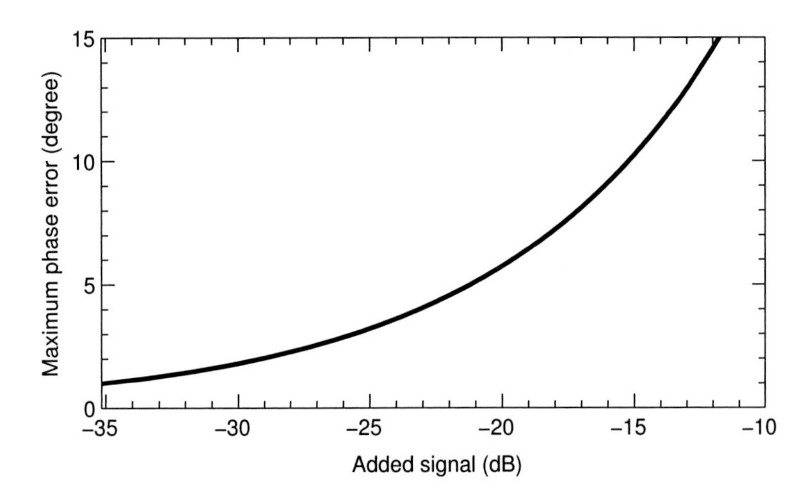


Figure 4.2: Maximum phase error due to an added disturbance signal.

The intrinsic characteristics of the channels must be determined in order to correct for the time delays and to equalize the signal level over the frequency band of operation. The cross-coupling must be eliminated from the measured data in order to determine the signal part caused by the imaged object. The mutual coupling, within either the Tx or Rx channels, influences the geometrical exactness of the phase centers of the antennas. This results in phase errors produced depending on the look angle, and hence degrade the focusing quality. This translates to the design requirement on the isolation between the RF channels of the same type. Fig. 4.2 shows the maximum error in the determination of the phase of a signal when a coupling signal of unknown phase is added to the first one. For coupling below $-35.2$ dB, the maximum phase error is rather less than $1^\circ$ and therefore has a negligible effect. In practice, such an isolation level is realizable [96, 97], and hence the mutual coupling effects are not considered in the system error correction in order to reduce the calibration complexity. The calibration of multistatic imaging arrays thus takes into account the correction for the intrinsic characteristics of each channel and the elimination of unwanted cross-coupling effects.

## 4.2 System error correction

The system error correction of the relevant effects discussed above requires a minimum of two calibration measurements. Considering the associated systematic errors illustrated in fig. 4.1, the measurement equation of the corresponding Tx and Rx channel at some frequency is

$$M = T \cdot (\Gamma_{obj} + \Gamma_{cc}), \tag{4.1}$$

where $T$ denotes the combined Tx and Rx channel transfer functions, $\Gamma_{obj}$ represents the actual reflection of the imaged object, and $\Gamma_{cc}$ includes all cross-coupling signals between the Tx and Rx channels. The transfer function $T$ includes the combined intrinsic characteristics of the Tx-Rx combination, i.e., the time delay and signal magnitude changes throughout the transmitting and receiving channels. It is physically defined at the phase centers of the Tx and Rx antennas. Therefore, it is practically impossible to measure the transfer function directly at its ports without facing further mutual coupling between the measuring setup and the structure of the antennas. Instead, it is more appropriate to deduce the value of $T$ without a real connection to the antennas, thus conserving an identical behavior of the antennas during calibration and measurement for imaging purpose. The cross-coupling part $\Gamma_{cc}$ describes generally all possible signal paths between the Tx and Rx channels, even before the antennas. It is only assumed that this signal portion does not get influenced by the imaged object, and hence seen to be static and can be systematically subtracted. Rearranging eq. 4.1 for $\Gamma_{obj}$ yields

$$\Gamma_{obj} = M/T - \Gamma_{cc}. \tag{4.2}$$

Eq. 4.2 includes two unknowns, i.e., $T$ and $\Gamma_{cc}$. These two unknowns can be solved by applying two measurement equations of known results, namely the calibration equations. By measuring any two objects of different and known reflection characteristics, the unknowns can be calculated. It is preferable to choose objects which produce significant difference in their complex reflection coefficients in order to increase the accuracy of the calibration procedure. This is easiest achieved in an imaging system by using match and offset-short measurements. The match measurement $M_m$ is simply realized by positioning an absorber material in the imaged volume, or by allowing for a large free space in front of the array in which the signal would practically scatter away from the array. The offset-short measurement $M_s$ is realized by positioning a highly reflective object in the imaged volume. Following the nomenclature used in calibrating the vector network analyzers, this standard is considered "offset" due to the spatial displacement between the actual reference plane and the position of the reflector. Assuming an ideal absorber, the measurement equations for the two calibration measurements are given by

$$M_m = T \cdot (0 + \Gamma_{cc}) \quad \text{and} \tag{4.3}$$

$$M_s = T \cdot (\Gamma_s + \Gamma_{cc}). \tag{4.4}$$

In order to solve eq. 4.3 and 4.4, the value of $\Gamma_s$ must be known or estimated. This can be made by either characterizing the reference object or by simulating it according to the known geometry. However the former would require measurement of the reference object exactly in the same geometrical arrangement, the latter allows for high flexibility in the choice of the geometrical arrangement and therefore is preferred. A sufficiently accurate electromagnetic simulation is thus essential to solve the equations. The applicability of such simulations will be addressed in the following sections. Accordingly, $\Gamma_s$ will be replaced with $\Gamma_{sim}$ in the following, in which $\Gamma_{sim}$ denotes the simulation result for the reflections of the reference reflector object calculated at the phase reference of the Tx and Rx antennas. Solving eq. 4.3 and 4.4 for $T$ and $\Gamma_{cc}$ thus yields

$$T = \frac{M_s - M_m}{\Gamma_{sim}} \quad \text{and} \tag{4.5}$$

$$\Gamma_{cc} = \frac{M_m}{M_s - M_m} \cdot \Gamma_{sim} \,. \tag{4.6}$$

## 4.3 Practical aspects

Several precautions should be considered in practice in order to ensure a calibration with an adequate quality for the imaging operation. The major factors influencing the accuracy of the calibration procedure are discussed next.

### 4.3.1 Signal-to-noise ratio

The calibration procedure involves collection of calibration measurements performed using different standards, i.e., at least two as discussed above. The collected measurements include noise. In ideal case, the noise is an additive white Gaussian noise of zero mean. As the noise is not deterministic in nature, the subtraction of the two subsequent calibration measurements, i.e., $M_s - M_m$, will result in a residual noise, which causes inaccuracy in the calculation of $T$ and $\Gamma_{cc}$ in eq. 4.5 and eq. 4.6, respectively. The application of eq. 4.2 requires a division with $T$, which would transform the zero mean noise to a non-zero mean one. This is harmful for the focusing process afterwards. Furthermore, the calibration measurement eq. 4.3 suffers from low SNR due to the absence of a strong reflector. Therefore, it is demanded to perform the calibration measurement with higher SNR than it is required in the imaging operation. For this, altering the hardware during calibration by using higher transmitting power level or narrower receiver bandwidths is not recommended, due to the fact that such changes in practice cause a consequent impact on the intrinsic characteristics of the channels and hence bring another source of inaccuracy for the calibration procedure. The better solution is to keep the system measurement procedure running the same way as during the imaging operation, and instead use numerical averaging techniques to integrate the received signal over time, and hence increase its SNR. Therefore, the data collected in eq. 4.3 and 4.4 is better averaged numerically in order to gain an increase in the SNR of the $M_s$ and $M_m$ values. The selection of the number of averaged measurements depends on the SNR available in a single measurement and the target SNR in the imaging results.

### 4.3.2 Cross-coupling

The signal $\Gamma_{cc}$ caused by the cross-coupling reduces the observed dynamic range of the receiver. Furthermore, having a relatively high value of $\Gamma_{cc}$ might push the receiver channels to get out of their linear operation range, and hence linear system error correction would fail. Therefore, care must be taken in the design of the array geometry in order to possibly reduce the cross-coupling. In this aspect, multistatic arrays are very advantageous, because they allow for large geometrical separation between the Tx and Rx channels, and hence offer the possibility for high isolation levels. The array introduced in sec. 2.9 demonstrates an example in which the number of geometrically close Tx-Rx combinations is purposely kept very low.

### 4.3.3 Multiple reflections

The calibration using two standards does not take into account any multiple reflections between the measured standard and the array itself. This is a critical issue for the calibration measurement of the reflector. On one hand, it should produce a high reflection level in order to ensure good SNR at the receiver; and on the other hand, it consequently raises multiple reflections with the array surface. The two conditions are almost contradictory and can be relaxed by the using of non-planar reflectors. The multiple reflection problem exists also during the imaging operation, in which the signal might bounce between the imaged object and the array surface. Assuming a well-calibrated array, the multiple reflections in this case can be suppressed by means of time domain gating. The proposed imaging method collects the measurements, however, in frequency domain rather than a direct time domain operation, and hence time domain representation can be made numerically by means of Fourier transformation. In order to reach the time domain representation of the reflected signals, the calibration must be free of multiple reflection effects in the first place.

### 4.3.4 Clutter

The calibration measurement eq. 4.3 and eq. 4.4 are in some sense idealized because of the absence of the influence of the surrounding objects. In practice, the imaging array integrates mechanical parts and the imaged volume includes unconsidered scatterers. This can range from the mechanical fixation of the system down to the floor covering under the scanning apparatus. The reflections of these objects, namely the clutter, cannot be described analytically, however being deterministic in nature. Consequently, numerical averaging used to reduce noise cannot help in this case. Instead, care must be taken in the preparation of the calibration geometry in order to minimize their effect by avoiding strong unconsidered reflective materials, e.g., metals, and geometries causing high reflections, e.g., planar surfaces or corner reflectors. It should be however noticed that, such objects which are constantly visible by the imaging array would produce repeatable signals during the two calibration measurements and hence will be illuminated as if being a cross-coupling effect.

### 4.3.5 Temperature variations

The Tx and Rx channels are realized using active devices, e.g., power amplifiers, low-noise amplifiers, buffers, and active mixers. The behavior of these components is generally temperature dependent, and hence change of their temperature would make the linear model describing them inaccurate. It is therefore a common practice when using radar systems, to allow them to reach a stable temperature before the actual operation. The same applies to the calibration of the imaging array. This extends to the consideration of the temperature cycle produced by each channel due to the repeated switching of the devices during the scans. It is therefore recommended to wait prior to the collection of the calibration measurements until the temperature of all channels is stable, or until a repeatable temperature cycling is reached in case of continuous switching.

In practice, the surrounding temperature of the imaging system varies over time, and therefore re-calibration must be taken into account. The two error terms of the multistatic array are not needed to be fully recalculated in order to perform a re-calibration. The cross-coupling term is mainly a function in the geometrical arrangement of the antennas and their connections, which are usually temperature stable. On the contrary, the transfer function is directly influenced by the active devices used in the front-ends, and hence is much dependent on the temperature. For this reason, re-calibration can be sufficiently made by repeating the measurement of the reference reflector and hence re-calculating the system error correction terms.

## 4.4 Sphere as a reflector

Fulfilling the calibration procedure discussed above requires the usage of a reflector with known characteristics. Flat reflectors like metal plates yield an inaccuracy in the calibration data due to the strong multiple reflections they produce. In practice, they are also not favorable due to the associated mechanical difficulties in the positioning and alignment of their position. They also have at least to be as large as the array surface itself, which is not feasible in many applications. Furthermore, the reflected signal power from a metal plate is much higher than from a typical imaged object. The hardware components in the Tx and Rx channels are often adjusted for power levels matching the case of the imaged object, and hence it is preferable to use a calibration object which results in a similar range of the reflected signal power. Making use of the high computation power available in the DBF systems, another type of a calibration object can be used to better fit the calibration requirements. For this, a metallic sphere is introduced. Using a sphere as a reference reflector has the following advantages.

- It ensures a reflection path between any Tx and Rx antenna.
- The curved surface strongly reduces the possibility for multiple reflections with the array surface.
- It eliminates the necessity for an exact mechanical alignment due to its full geometrical symmetry.
- Edge scattering effects are fully avoided.
- The reflected signal is mainly co-polarized and hence is well-suited for co-polarized arrays.
- It relaxes the necessity for mechanical removal after the calibration procedure, as it could be left in the shadow region behind the scanned object, thus allows for a self-calibration when the scanned volume is empty.
- In comparison to simple metal plate reflectors, it is much lighter and less costly.

### 4.4.1 Problem definition

The geometry of the problem as well as the degree of accuracy necessary in the solution, form together the frame of validity for the possible solution method. The considered problem is a bistatic scattering problem of an assumed perfectly conducting sphere. The bistatic angle reaches a large value, which could be close to 120°, thus describing the angle between the two lines connecting the transmitter and receiver to the sphere center. The sphere is located in relatively close range to the transmitter and receiver, and hence the curvature of the phase front is highly spherical. The sphere diameter is chosen much larger than the incident wavelength, with at least one order of magnitude. The demanded accuracy in the calculation of the phase of the reflection coefficient is considered sufficient to reach 1°. This value is appropriate for the proposed imaging method, in which the processing of the phase information and the focusing of the image is practically not affected by such small phase error. This translates to a demanded accuracy of $\lambda/360$ in the determination of the geometrical signal paths. In accordance to the imaging method used, the demand on accuracy in the determination of the magnitude of the reflection coefficient is much relaxed. In practice, errors of approximately 1% in signal power are fully negligible. The applying of the sphere as a calibration object is constrained to the co-polarized arrays, in which the transmitter and receiver antennas use same polarization. This is due to the dominant co-polarized scattering the sphere produces, as proved later on.

### 4.4.2 Known solutions

The scattering problem of the sphere was addressed in various applications. Next, an overview of the known approaches is given.

#### 4.4.2.1 Analytical method

One of the early investigations made to solve the scattering problem of conductive spheres was made by Gustav Mie, famously known as Mie solution. Mie solution was published in 1908 [98] in German, and was afterwards translated in English and also expanded, as in the translation of Newman [99] in 1978, and the work of Goodrich et al. [100] in 1961. The latter expanded the discussion on the solutions of the problem using physical optics and numerical techniques. Because the work of Mie was originally targeting the optical domain, the emphasis was made for the far-field solution using incident plane waves. Mie stated in his work [98] on page 393:

> *"Das vorliegende Problem ist als gelöst zu betrachten, sobald es gelingt, die ebene Welle, die auf die Kugel auftrifft, in Ausdrüke ... zu zerlegen".*

According to Newman [99], it translates to: *"The present problem can be considered solved as soon as the plane waves incident on the sphere are resolved into expressions ..."*. Therefore, the accurate solution given

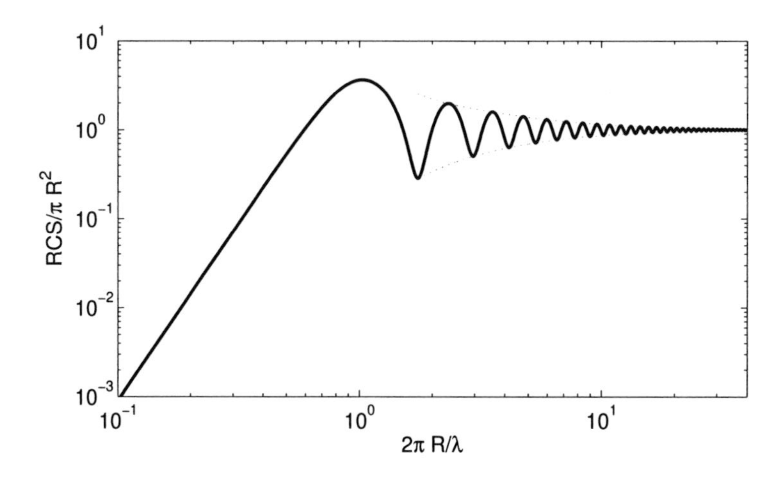


Figure 4.3: Mie solution for monostatic RCS of a perfect conductive sphere, showing the asymptotic behavior for electrically large spheres.

by Mie is strictly not applicable when the wavefront is not plane. For practical considerations, the size of the calibration sphere must be large in order to ensure sufficient signal level over thermal noise. The sphere is also placed in some close range to the array surface making the curvature of the wavefront relatively high. Therefore, the incident wave must be modeled by a spherical wavefront, making the direct application of Mie solution unsuitable to the proposed geometry. In Mie solution for conductive spheres, it is however readable that when the sphere radius is large relative to the incident wavelength, the reflection takes place more locally on the sphere surface and the influence of creeping waves asymptotically vanishes. The scattering of radar objects in far-field is often described by the radar cross section (RCS) based on looking angle and frequency used. The asymptotic behavior of the scattering of the sphere is well illustrated by the monostatic radar cross section, which can be directly numerically calculated from the Mie solution. Based on the formulation in [101], the normalized RCS of a perfectly conductive sphere of a radius $R$ is calculated as

$$\frac{\mathrm{RCS}}{\pi R^2} = \left(\frac{\lambda}{2\pi R}\right)^2 \cdot \left|\sum_{n=1}^{\infty} \frac{(-1)^n \cdot (2n+1)}{H_n^{(2)\prime}(2\pi R/\lambda) \cdot H_n^{(2)}(2\pi R/\lambda)}\right|^2 , \tag{4.7}$$

where $H$ denotes the spherical Hankel function. Fig. 4.3 illustrates the above solution using a truncated sum to $n = 50$.

Referring to the numerical results of the Mie solution in fig. 4.3, the fluctuations in the RCS vanishes rapidly while increasing the ratio

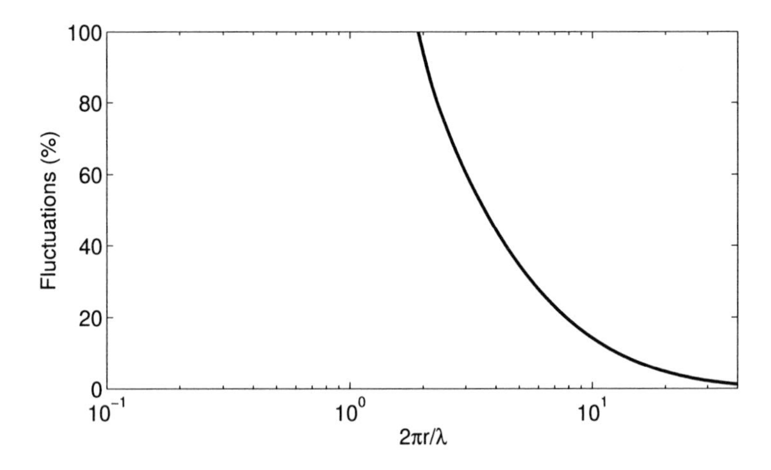


Figure 4.4: Fluctuations of the normalized RCS shown in fig. 4.3, which vanishes rapidly for electrically large spheres.

$2\pi R/\lambda$, as illustrated in fig. 4.4. The fluctuations are approximately 1% for $2\pi R/\lambda = 200$. For instance, at 100 GHz it suggests the usage of a sphere radius of at least 10 cm in order to ensure local scattering behavior. This leads also to the fact that the reflection coefficient of a hemisphere will not alter noticeably from the full sphere case, as long as the scattering of the edges of the hemisphere does not contribute significantly to the scattered field. Same can be said about the creeping waves, leading to a significant simplification in the demanded solution.

#### 4.4.2.2 Earth reflections

Bistatic reflection on the sphere surface was often addressed in literature for the solution of multipath problems in atmospheric communication links and some radar applications. Having two antennas over ground, a direct line-of-sight path and a reflection over ground path interfere at the receiver. Therefore, the calculation of the exact path difference between the two geometrical paths, and accordingly the determination of the specular reflection point, is essential for the design of the communication link. In spite of the strong similarity of the problem, the known solutions either make use of the fact that the radius of the earth is much larger than the antenna heights, as in [102], or assume small bistatic angles as in [103]. Both assumptions are not applicable for the considered problem here, where the calibration sphere is considerably small and the bistatic angles are relatively large.

#### 4.4.2.3 Computational electromagnetics

Full-wave solvers could be used to simulate the scattering of the sphere. Due to the large separation space between the imaging array and the sphere, numerical solvers using a full meshing of the simulated space cannot be used effectively here. The problem in this case is electrically huge, leading to memory requirements, which are not feasible even with modern computers. Instead, the numerical methods requiring a meshing of the object surface are better suited for this scattering problem. The most accurate would be the method of moments. Although, the computation power needed for it could be made available with modern computers, the associated cost is quite high and the simulation time is very long. All this makes the usage of full-wave solvers unsuitable for this task.

Simulation methods relying on high frequency approximations are more applicable in this case. Physical optics (PO) could be used, which discretizes the illuminated surface of the reflecting sphere, then calculates the surface currents excited by the incident field, and afterwards integrates the radiated field out of each surface portion at the target receiver. The simulation would deliver sufficient accuracy as needed for the imaging application, however it still requires relatively high computation power and memory resources for electrically large spheres. Therefore, it will be considered next as a verification method rather than a simulation tool during the calibration procedure. Geometrical optics (GO) is also suitable to solve this scattering problem, where the computational power is considerably reduced without a significant loss of accuracy. Therefore, a dedicated GO solver for this problem is introduced in sec. 4.4.3.

In order to illustrate the scattering behavior of electrically large spheres, the following simulation is established using the PO method of the commercial solver FEKO [104]. A sphere of 15 cm radius is positioned at 1 m away from an ideal dipole. The dipole is polarized in $x$-direction and transmits an 80 GHz signal. The geometrical arrangement is illustrated in fig. 4.5. Fig. 4.6 shows the simulation results for the $x$-component of the reflected field. The magnitude is normalized to the transmitted field strength and displayed in dB. The small fluctuations in the magnitude are easily seen. The reduction in the field strength along the $x$-direction is higher than for the $y$-direction due to the orientation of the dipole.

#### 4.4.2.4 Paraxial optics approach

The scattering problem from a perfectly conductive sphere can be understood as a mirroring problem. The sphere itself can be interpreted as a convex spherical mirror, which reflects a wavefront corresponding to a virtual mirrored transmitter inside the sphere, the same way it works in optics. This approach follows the rules of paraxial optics, in which the

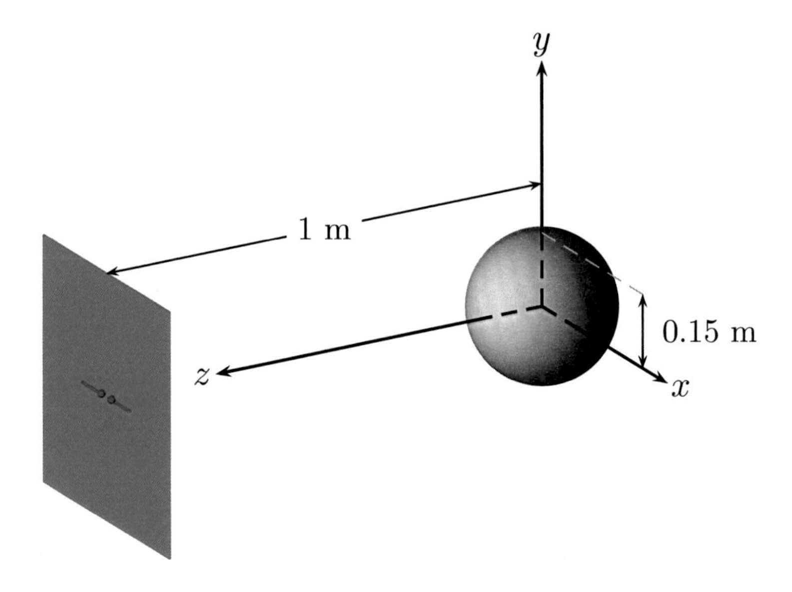


Figure 4.5: The geometrical illustration of the sphere simulation with an ideal dipole antenna polarized in $x$-direction. The dipole is positioned on the $z$-axis at one meter distance to the sphere center. The indicated plane shows the area where the scattered field is calculated. The plane is perpendicular to the $z$-axis; and is centered to the dipole.

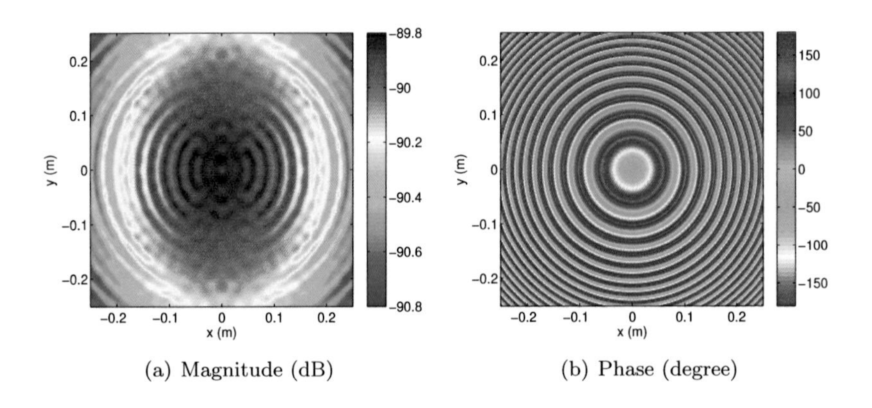


(a) Magnitude (dB)

(b) Phase (degree)

Figure 4.6: Simulation results using PO method at 80 GHz showing the $x$-component of the reflected wavefront according to the geometry indicated in fig. 4.5.

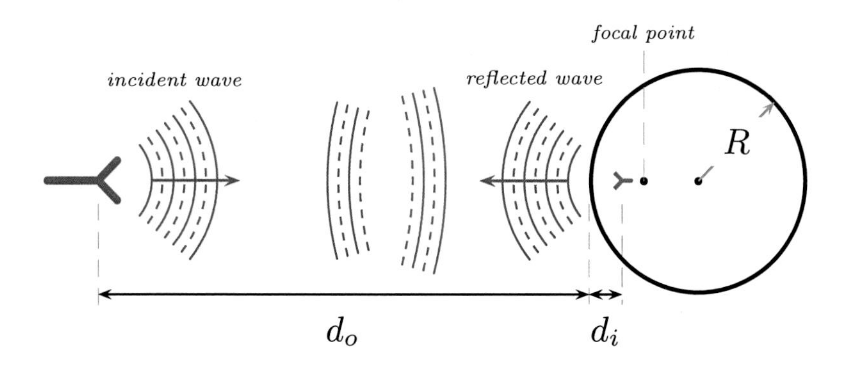


Figure 4.7: The geometrical illustration of the sphere when considered as a spherical mirror.

virtual position of the transmitter image is only valid within a small angle variation around the optical axis. In other words, the approximation does not hold true for large variations in the bistatic angle, as it is the case with close range operation. It is, however, useful to consider it as a simplification of the simulation task.

For a spherical mirror of radius $R$, the focal point is at $R/2$ within the sphere [105]. This results in the mirror formula

$$\frac{1}{d_i} = \frac{1}{d_o} + \frac{2}{R} \,, \tag{4.8}$$

where $d_o$ and $d_i$ denote the distance from the mirror surface to the object and to its virtual image, respectively. Fig. 4.7 illustrates the geometrical arrangement and the mirroring of an example transmitter.

For each transmitter, a virtual transmitter image can be found in the sphere, from which a reflected wavefront to all receivers can be determined. Nevertheless, this process lacks: 1) the dependency on the bistatic angle, as simplified by paraxial methods, and 2) the influence of signal divergence and hence the change in the power density. Both require the use of a more consolidated simulation method. Fig. 4.8 illustrates the phase difference between the simulated wavefronts using paraxial method and the PO simulation presented above. The result shows the progressive phase error for angles far from the optical axis.

### 4.4.3 Geometrical optics model

The determination of the sphere reflection, in accordance to the required geometrical arrangement discussed above, is essential for the success of the calibration procedure. The required accuracy and the importance of fast

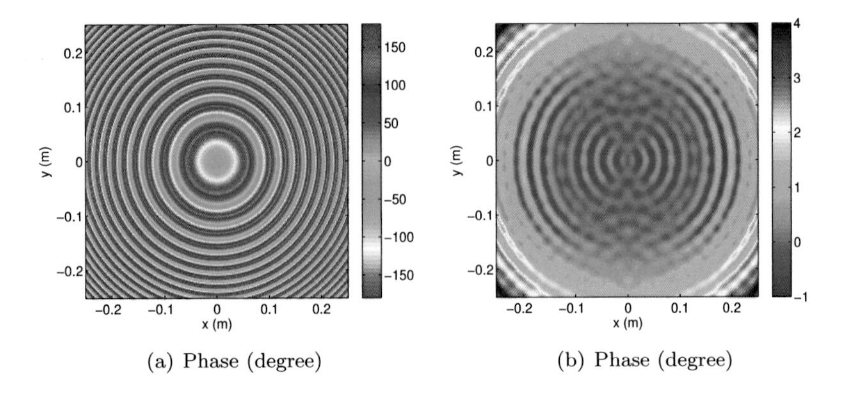


(a) Phase (degree) (b) Phase (degree)

Figure 4.8: On the left, the simulation result of the reflected wavefront when considering the sphere as a convex mirror. On the right, the normalized phase error when compared to the PO result in fig. 4.6(b) is shown, indicating the progressive phase error for angles far from the optical axis.

simulations suggest the use of a dedicated numerical solution based on high-frequency approximations. Numerical methods using integrations, e.g., PO, require a high computational power and often a long calculation time, making them also unsuitable. Having a calibration object without hard edges and with a radius of curvature much larger than the used wavelengths, makes the use of geometrical optics very suitable to the problem [57]. Hereafter, a dedicated GO solver for the simulation of the scattering of an electrically large perfect conducting sphere is introduced. Care is taken in the calculation of the phase factor, which plays a major role in imaging reconstruction. This leads to the importance of the accurate determination of the specular reflection point on the sphere surface. The GO solver takes into consideration the antenna patterns, the free space attenuation, the signal diversion on the sphere surface, and the change of polarization at the receiver. All together lead to a sufficiently accurate calculation of the sphere scattering, which will be verified later by simulation comparisons with the PO method as well as experimentally. The solver assumes a homogeneous medium, e.g., air, and a single reflecting sphere placed many wavelengths away from the transmitter and receiver antennas in order to ensure far-field condition relative to the antennas. The receiver should not be in the shadow region of the transmitter, which is the case for arrays used in reflection imaging.

The purpose of the simulation is to determine the complex reflection coefficient $\Gamma_{sim}$, with the reference planes defined at the phase centers of the transmitter and receiver antennas. The complex reflection coefficient is described by a phase and a magnitude factor, given as

$$\Gamma_{sim} = A \cdot e^{j\phi}. \tag{4.9}$$

The phase factor $\phi$ corresponds to the signal delay, and hence depends on the geometrical path to the sphere and back. The magnitude factor $A$ depends on the space attenuation and the characteristics of antennas.

#### 4.4.3.1 The phase factor

The geometrical optics uses the concept of rays, in which the signal is assumed to take a straight path, i.e., light ray, in the medium until being interrupted by another medium. At the receiver, only a certain bundle of rays will be received depending on the underlying geometrical arrangement. These rays are the ones exhibiting a geometrical variation minimum, either globally or locally. In other words, when several reflection scenarios exist, many paths can lead to the receiver; and each of them is the one delivering minimum geometrical path in its local neighborhood. This is addressed by the Fermat's principle as in [57]. Under the condition of a single reflection in a homogeneous medium, Fermat's principle[§] leads to the famous law of reflection stating that the angle of incident and reflection must be equal. This is the condition resulting in the minimum signal path, and accordingly the one defining the phase delay between the transmitter and the receiver. The specular point is hence defined to be the geometrical solution on the reflecting surface at which the angle of incident and reflection are equal. The scattering process is assumed to be local at this point. Therefore, the accurate finding of this point is necessary for the geometrical optics solution.

Without loss of generality, the sphere center is assumed to be the coordinates origin, and the position of the phase centers of the transmitter and the receiver antennas lay in the ($z = z_a$)-plane. The sphere is of a perfectly conducting surface and has a radius $R$. The geometry is illustrated in fig. 4.9, in which the transmitter position vector $\mathbf{t}$, the receiver position vector $\mathbf{r}$, the position vector $\mathbf{s}$ at the specular point are indicated. For $\mathbf{p}$ describing a general point on the sphere surface, which satisfies

$$\|\mathbf{p}\| = R, \tag{4.10}$$

the Fermat's principle for the specular point $\mathbf{s}$ can be written as

$$\mathbf{s} = \mathbf{p} \text{ for } (\|\mathbf{t} - \mathbf{p}\| + \|\mathbf{r} - \mathbf{p}\|) \text{ is minimum.} \tag{4.11}$$

[§]It was introduced by Pierre de Fermat in 1662 as a principle in optics domain, which also holds true at longer wavelengths. Now it is also known as principle of least time, and is proved and generalized in many contributions as in [57].

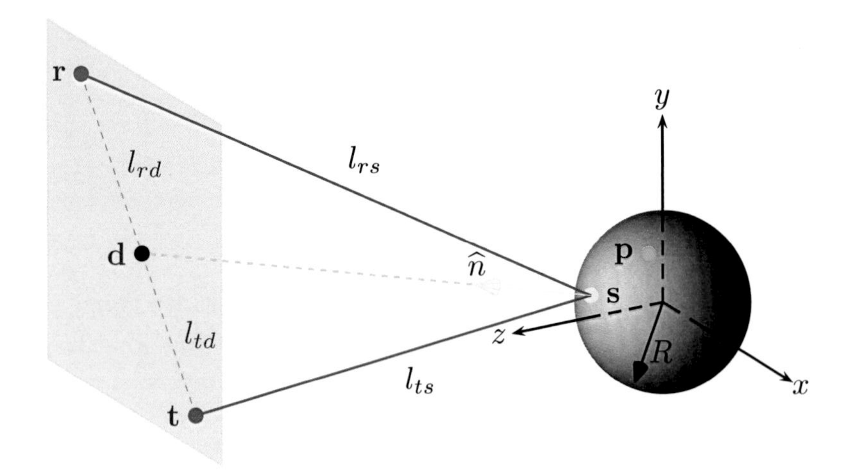


Figure 4.9: Geometrical description used in the GO solver indicating the position of the reflecting sphere, the positions of the transmitter and the receiver, and the quantities used in the determination of the phase factor.

Alternatively, this might be expressed as

$$\mathbf{s} = \mathbf{p} \text{ when } \frac{\partial}{\partial \mathbf{p}} \left( \|\mathbf{t} - \mathbf{p}\| + \|\mathbf{r} - \mathbf{p}\| \right) = 0\,, \tag{4.12}$$

which can result in two solutions for the sphere front and back sides, however the latter is not a valid solution for the problem here and can be ignored.

Finding the specular point using eq. 4.11 or eq. 4.12 is numerically costly, as it involves a minimization problem in three dimensions. It can be easily reduced to a two-dimensional problem by using the constraint in eq. 4.10, however still a time-consuming process. Due to the necessity to find a solution to each transmitter-receiver combination in the imaging array, a direct analytical solution to the geometrical problem is highly desired. Although the problem can be reduced to a quartic equation [106, 107], which is algebraically solvable§, the analytical solution is very lengthy and can become numerically unstable. Instead, several approaches were introduced in literature [102, 103, 109] to solve the problem depending on the simplifications feasible to the target application. Therefore, a dedicated approximate analytical geometrical solution is presented next, which can yield to a sufficiently accurate determination of the specular point suitable to the required frame of accuracy.

§This problem attracted many scientists over centuries and is named historically as Alhazen's problem [108] after the Arab scholar Al-Hasan ibn Al-Haitham.

Referring to fig. 4.9, the point $\mathbf{d}$ denotes the intersection between the line connecting the transmitter position $\mathbf{t}$ and the receiver position $\mathbf{r}$ with the line connecting the origin and the specular point $\mathbf{s}$. The latter line is perpendicular to the sphere surface at $\mathbf{s}$, as it goes through the sphere center, i.e., the origin. Knowing that the incident ray, the reflected ray, and the line perpendicular to the surface all lay in the same plane, i.e., the plane of incidence, then the intersect point $\mathbf{d}$ must exist. This point is used to assist in formulating the geometrical solution. Additionally, the distances to this point drawn from $\mathbf{t}$ and $\mathbf{r}$ are named $l_{td}$ and $l_{rd}$, respectively. Similarly, the distances to the specular point $\mathbf{s}$ are named $l_{ts}$ and $l_{rs}$. According to the angle bisector theorem, for equal angle of incidence $\theta_t$ and angle of reflection $\theta_r$, the distances are related by

$$\frac{l_{td}}{l_{rd}} = \frac{l_{ts}}{l_{rs}} . \tag{4.13}$$

The position of $\mathbf{d}$ is always located between $\mathbf{t}$ and $\mathbf{r}$, and can therefore be described as

$$\mathbf{d} = \mathbf{t} + (\mathbf{r} - \mathbf{t}) \cdot \frac{l_{td}}{l_{tr}} , \tag{4.14}$$

where $l_{tr}$ is the distance between $\mathbf{t}$ and $\mathbf{r}$. After applying eq. 4.13 in eq. 4.14, the position of $\mathbf{d}$ is expressed as

$$\mathbf{d} = \mathbf{t} + (\mathbf{r} - \mathbf{t}) \cdot \left( \frac{1}{1 + l_{rs}/l_{ts}} \right) . \tag{4.15}$$

Once the point $\mathbf{d}$ is determined, the position of $\mathbf{s}$ is easily found by scaling the vector length as

$$\mathbf{s} = R \cdot \frac{\mathbf{d}}{\|\mathbf{d}\|} . \tag{4.16}$$

According to the formulations in eq. 4.15 and eq. 4.16, the solution of the geometrical problem depends on the ratio $l_{rs}/l_{ts}$. In the monostatic case, i.e., $\mathbf{t} = \mathbf{r}$, the ratio is equal unity leading to the trivial solution of $\mathbf{d} = \mathbf{t} = \mathbf{r}$. In the bistatic case, this ratio is not easily calculated and is, instead, approximated or iteratively found. In order to reach a closed-form solution to the problem, the following approximation is applied

$$\frac{l_{rs}}{l_{ts}} \approx \frac{\|\mathbf{r}\|}{\|\mathbf{t}\|} . \tag{4.17}$$

In this approximation, the ratio is directly related to the positions of the transmitter and the receiver, however the radius of the sphere is ignored. Therefore, it is most accurate for the case when

$$\|\mathbf{t}\| \gg R \text{ and } \|\mathbf{r}\| \gg R . \tag{4.18}$$

These strictly apply in the far-field case, where $l_{rs}/l_{ts} \approx 1$ and $\mathbf{d}$ is located at $(\mathbf{t}+\mathbf{r})/2$. The approximation, however, will be proven later to be sufficiently accurate for the considered close range case by using comparisons with PO simulations as well as experimental verifications. Finally, the approximate solution of $\mathbf{d}$ is written from eq. 4.15 and eq. 4.17 as

$$\mathbf{d} \approx \mathbf{t} + (\mathbf{r} - \mathbf{t}) \cdot \left( \frac{\|\mathbf{t}\|}{\|\mathbf{t}\| + \|\mathbf{r}\|} \right) . \tag{4.19}$$

After the determination of $\mathbf{d}$, the specular point $\mathbf{s}$ is calculated from eq. 4.16, then $l_{ts}$ and $l_{rs}$ are also known. According to the boundary conditions for reflection on perfectly electric conductive surfaces, a phase lag of $\pi$ is considered [84]. Therefore, the phase factor at the wavelength $\lambda$ is hence equal to the phase delay along the geometrical path subtracted from $\pi$. Thus the total phase factor is given by

$$\phi = -\frac{2\pi}{\lambda} \cdot (l_{ts} + l_{rs}) - \pi . \tag{4.20}$$

#### 4.4.3.2 The magnitude factor

In order to determine the magnitude factor of the reflection coefficient, the signal path is traced from the transmitter to the receiver. For this, the considered signal path is the one found by the solution discussed above; and the scattering of the signal is assumed to be local at the specular point $\mathbf{s}$. The gain of the transmitter and receiver antennas pointing towards the specular point are considered known to the simulator, either from a theoretical model, an empirical model, or a full-wave simulation of the antennas. Any losses or impedance mismatches in the antennas are hence included in the gain values. Usually, the antennas utilized in imaging systems exhibit identical characteristics; and hence it is sufficient to characterize one of them and make it available to the simulator. In some applications, the type of the transmitter and receiver antennas might differ. In such a case, the different antenna gains have to be taken into account.

The signal emitted from the transmitting antenna diverges in space and part of it reaches the sphere surface. Due to the curvature of the sphere, the reflected signal back towards the receiver antenna includes further diversion. At the receiver antenna, the incident signal would fit partially with the polarization of the antenna, and hence only certain part of the signal power will be polarization-matched to the receiver. At last, the antenna will receive part of the signal power according to its gain at the corresponding incident angle. For a quantitative description of the magnitude factor, these effects are formulated next in a closed-form.

Fig. 4.10 illustrates the geometrical arrangement similarly to fig. 4.9, however with the emphasis on the quantities used in the calculations of the magnitude factor. The unit vectors are indicated with a $\widehat{\cdot}$ symbol. The transmitter and receiver antennas are assumed co-polarized to $\widehat{E}_{ant}$. $\widehat{E}_{ts}$ and $\widehat{E}_{rs}$ denote the polarization unit vectors at the specular point before and after the reflection, respectively. The positions $\mathbf{t}$, $\mathbf{r}$, $\mathbf{s}$, and the origin are in the same plane, i.e., the plane of incidence, for the reasons discussed before. Considering far-field radiation for the antennas, transverse electromagnetic mode (TEM) is the propagation mode. Therefore, $\widehat{E}_{ant}$ and $\widehat{E}_{ts}$ are in the same plane; and $\widehat{E}_{ts}$ is perpendicular to the line $(\mathbf{t} - \mathbf{s})$. Similarly, $\widehat{E}_{rs}$ is perpendicular to the line $(\mathbf{r} - \mathbf{s})$, because the reflected wave received by the receiver antenna is in TEM mode.

For the sake of simplicity, the square of the magnitude factor is used, which relates the received power $P_r$ to the transmitter power $P_t$ directly through

$$A^2 = \frac{P_r}{P_t}. \tag{4.21}$$

Using the free space attenuation relation [110], the power density available at the specular point prior to the reflection, is expressed as

$$\frac{P_t \cdot G_{ts}}{4\pi \cdot l_{ts}^2}, \tag{4.22}$$

where $G_{ts}$ denotes the transmitter antenna gain value towards the specular point $\mathbf{s}$. The signal is then reflected from the surface with a radius of curvature less than the incident signal. The center of this curvature lies in a virtual point within the sphere, and at a distance $\rho$ from the surface, as illustrated in fig. 4.11. This type of interaction between an incident wave and a curved surface had been studied in [77], and proved the following geometrical relation

$$\frac{1}{\rho} = \frac{1}{l_{ts}} + \frac{2}{R\cos(\theta_t)}, \tag{4.23}$$

where $\theta_t$ is the angle of incidence[§]. Having a sphere as a reflector, the relation is applicable to the two principle directions perpendicular to the incident ray[¶]. Rearranging eq. 4.23, $\rho$ is given by

$$\rho = \frac{R\cos(\theta_t) \cdot l_{ts}}{R\cos(\theta_t) + 2l_{ts}}. \tag{4.24}$$

The power density of the reflected signal depends on the new radius of curvature, i.e., $\rho + l_{rs}$. In order to get the power density of the reflected

[§] It is noticeable here that this relation converges to eq. 4.8 under paraxial condition, i.e., small values of $\theta_t$.

[¶] A little approximation is considered here, as for large values of $\theta_t$ the signal diversion behavior would not be identical in the two directions. However, this is out of the scope of the required accuracy in this analysis.

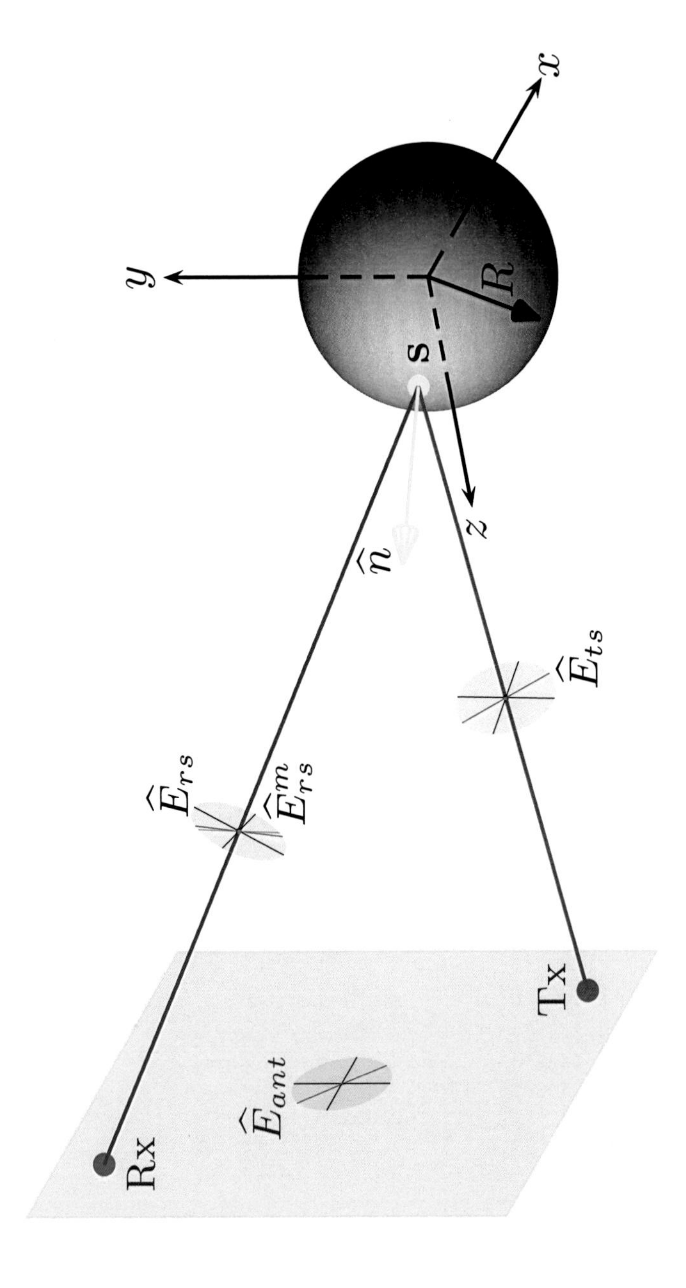


Figure 4.10: Geometrical description used in the GO solver for the determination of the magnitude factor.

signal, two hypothetical areas perpendicular to the reflected signal path are considered. The first one $a_1$ is placed infinitely close to the specular point including its neighborhood. The second one $a_2$ is placed along the reflected signal path including all rays passed through the first one. Accordingly, the power radiating through both areas must be equal. Considering that the first area has a radius of curvature equal to $\rho$ and the second one equal to $\rho + l_{rs}$, thus the power densities at the two areas are related by

$$P_1 \cdot a_1 = P_2 \cdot a_2 \Longrightarrow \frac{P_2}{P_1} = \frac{a_1}{a_2} = \left(\frac{\rho}{\rho + l_{rs}}\right)^2 . \tag{4.25}$$

Consequently, the power density of the reflected signal is expressed as

$$\frac{P_t \cdot G_{ts}}{4\pi \cdot l_{ts}^2} \cdot \left(\frac{\rho}{\rho + l_{rs}}\right)^2 . \tag{4.26}$$

The reflected signal is in general partially co-polarized with the receiver antenna. This polarization mismatch can be deduced from the known geometry by applying boundary conditions at the specular point. The normal vector $\widehat{n}$ at the specular point is easily given by

$$\widehat{n} = \frac{\mathbf{s}}{\|\mathbf{s}\|} . \tag{4.27}$$

As the reflection is considered local at the specular point, boundary conditions for planar surfaces with infinite electric conductivity apply, in which the tangential component of the electric field must vanish [84]. Thus, at $\mathbf{s}$, the incident field $E_{ts}$ and the reflected one $E_{rs}$ must satisfy

$$\widehat{n} \times (E_{ts} + E_{rs}) = 0 . \tag{4.28}$$

Considering that $E_{ts}$ and $E_{rs}$ are TEM waves [111], applying eq. 4.28 requires that their tangential (symbol $.^{\|}$) and perpendicular (symbol $.^{\perp}$) components relative to the surface follow

$$E_{rs}^{\|} = -E_{ts}^{\|} \quad \text{and} \quad E_{rs}^{\perp} = +E_{ts}^{\perp} . \tag{4.29}$$

Hence, it is evident that at $\mathbf{s}$

$$E_{rs} + E_{ts} = 2\,(E_{ts} \cdot \widehat{n})\,\widehat{n} . \tag{4.30}$$

It is convenient to use unit vectors in the description of the polarization. Eq. 4.30 is therefore rewritten for $\widehat{E}_{rs}$, as

$$\widehat{E}_{rs} = 2\left(\widehat{E}_{ts} \cdot \widehat{n}\right)\widehat{n} - \widehat{E}_{ts} . \tag{4.31}$$

The polarization vector $\widehat{E}_{ts}$ is related to the polarization $\widehat{E}_{ant}$ and the looking direction $\widehat{r}_{ts}$ of the transmitting antenna, defined as a unit vector pointing from $\mathbf{t}$ to $\mathbf{s}$, namely

$$\widehat{r}_{ts} = \frac{\mathbf{s}-\mathbf{t}}{\|\mathbf{s}-\mathbf{t}\|} \quad . \tag{4.32}$$

Similarly, $\widehat{r}_{rs}$ is a unit vector pointing from $\mathbf{r}$ to $\mathbf{s}$. For TEM wave, the following description for the signal polarization of the transmitter holds

$$\widehat{E}_{ts} = \frac{\widehat{E}_{ant} - \left(\widehat{E}_{ant}\cdot\widehat{r}_{ts}\right)\widehat{r}_{ts}}{\left\|\widehat{E}_{ant} - \left(\widehat{E}_{ant}\cdot\widehat{r}_{ts}\right)\widehat{r}_{ts}\right\|} \quad . \tag{4.33}$$

By substituting eq. 4.33 into eq. 4.31, the value of the polarization vector of the reflected signal is reached. Knowing the polarization vector $\widehat{E}_{ant}$ of the receiver antenna[§], the polarization-matched portion of the signal can now be calculated. First, the projection of $\widehat{E}_{ant}$ on the plane perpendicular to $\widehat{r}_{rs}$ is determined, which corresponds to the fully polarization-matched case. The unit vector is thus named $\widehat{E}^{m}_{rs}$ and is given as

$$\widehat{E}^{m}_{rs} = \frac{\widehat{E}_{ant} - \left(\widehat{E}_{ant}\cdot\widehat{r}_{rs}\right)\widehat{r}_{rs}}{\left\|\widehat{E}_{ant} - \left(\widehat{E}_{ant}\cdot\widehat{r}_{rs}\right)\widehat{r}_{rs}\right\|} \quad . \tag{4.34}$$

Then, the polarization-matched power density is calculated by introducing the term $\cos^2\left(\theta_{pol}\right)$ to the power relation in eq. 4.26. The $\theta_{pol}$ denotes the angle between the polarization vectors $\widehat{E}^{m}_{rs}$ and $\widehat{E}_{rs}$, as illustrated in fig. 4.12. This term can be expressed in vector notation as

$$\cos^2\left(\theta_{pol}\right) = \left(\widehat{E}^{m}_{rs}\cdot\widehat{E}_{rs}\right)^2 \quad . \tag{4.35}$$

Lastly, the received power by the receiver antenna from the polarization-matched power density is found by the multiplication with the receiver effective aperture [110], namely

$$\frac{\lambda^2}{4\pi}G_{rs} \quad , \tag{4.36}$$

where $G_{rs}$ denotes the receiver antenna gain value towards the specular point. The magnitude factor at the wavelength $\lambda$ can now be expressed

[§] If the receiver antenna assumes different polarization than the transmitter antenna, then the corresponding polarization should be used here.

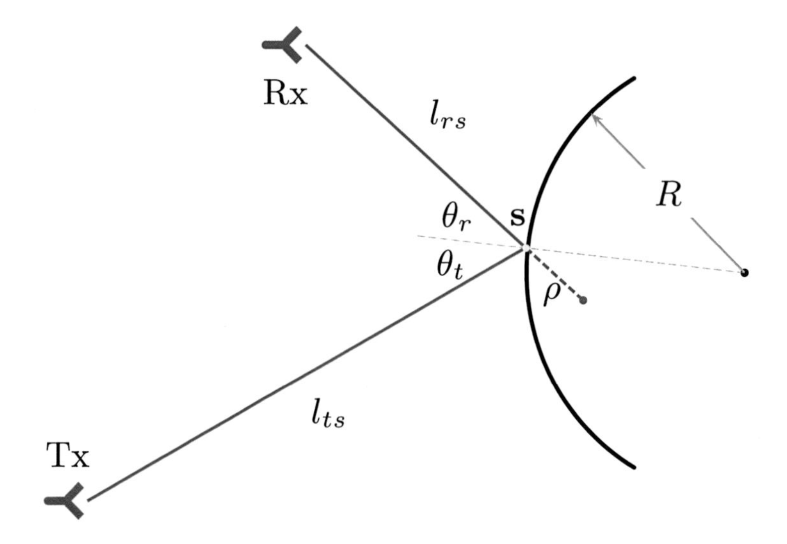


Figure 4.11: Detailed view at the specular point.

by

$$A^2 = \frac{P_r}{P_t} = \underbrace{\underbrace{\underbrace{\left(\frac{G_{ts}}{4\pi \cdot l_{ts}^2}\right)}_{\text{power density at } \mathbf{s}} \cdot \underbrace{\left(\frac{\rho}{\rho + l_{rs}}\right)^2}_{\text{diversion effect}}}_{\text{power density at receiver}} \cdot \underbrace{\cos^2\left(\theta_{pol}\right)}_{\text{pol. matching}}}_{\text{polarization-matched power density at receiver}} \cdot \left(\frac{\lambda^2}{4\pi} G_{rs}\right) . \tag{4.37}$$

It is worth mentioning that the results in eq. 4.20 and eq 4.37 successfully satisfy the reciprocity theorem [84].

#### 4.4.3.3 Verification simulations

The simulation results of the GO solver are now compared with the PO results presented in sec. 4.4.2.3, which were simulated using an 80 GHz excitation. Following the same geometry as in fig. 4.5, the simulation is repeated using the GO solver and the resulting $x$-component of the reflected field is computed and presented in fig. 4.13. The magnitude and phase responses highly agree with the PO results, with the exception of the omitted fluctuations discussed in sec. 4.4.2.1. Fig. 4.14 shows a comparison between the PO and the GO results of fig. 4.6 and fig. 4.13, respectively. The comparison is made by considering the ratio of the two complex reflection coefficients. The ratio between the two solutions is in

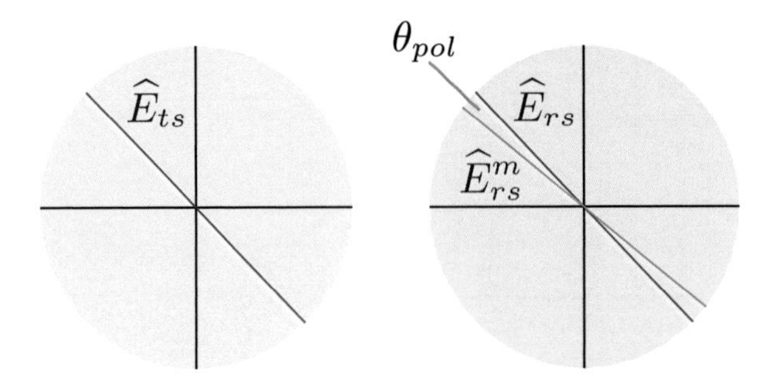


Figure 4.12: Illustration of the change in the state of polarization for the case in fig. 4.10.

average below 0.1 dB in magnitude and 0.5° in phase. The magnitude shows overshoots at the positions of the fluctuations maxima and minima as expected. A further verification comparison is made for the case of a transmitter positioned at $(x, y, z)$ position of $(0.25, 0.25, 1)$ m, in order to examine the case for a Tx antenna placed at the corner of the aperture. The simulation results for the PO and the GO solvers are illustrated in fig. 4.15 and fig. 4.16, respectively. The comparison between the two solutions is illustrated in fig. 4.17, hence again indicating good agreement.

The results in the two verification simulations prove the GO solver to compute the reflection coefficients with the accuracy required for the imaging operation, as discussed in sec. 4.4.1. The approximation made in eq. 4.23 is responsible for the little deviation in the magnitude shown in fig. 4.17(a), however, the error is not significant and is ignored.

As discussed before, the Rx antenna receives the polarization-matched portion of the reflected field. In order to illustrate the effect of the polarization mismatching presented in eq. 4.35, the values of $\theta_{pol}$ and the corresponding values of $\cos^2(\theta_{pol})$ are visualized for the geometry arrangement of fig. 4.5. Fig. 4.18 presents the computation results, thus showing the progressive polarization mismatching towards the corners of the aperture. It clearly indicates that the bistatic scattering of electrically large spheres is perfectly co-polarized in the E-plane and H-plane of the Tx antenna. Apart of this, the scattered signal is partially co-polarized.

#### 4.4.3.4 Verification measurements

The complex reflection coefficients of a test metallic sphere are verified using three scans. Mechanical scans collecting the reflected signals are performed for the three Tx positions illustrated in fig. 4.19, namely for: 1) upper transmitter, 2) central transmitter, and 3) corner transmitter.

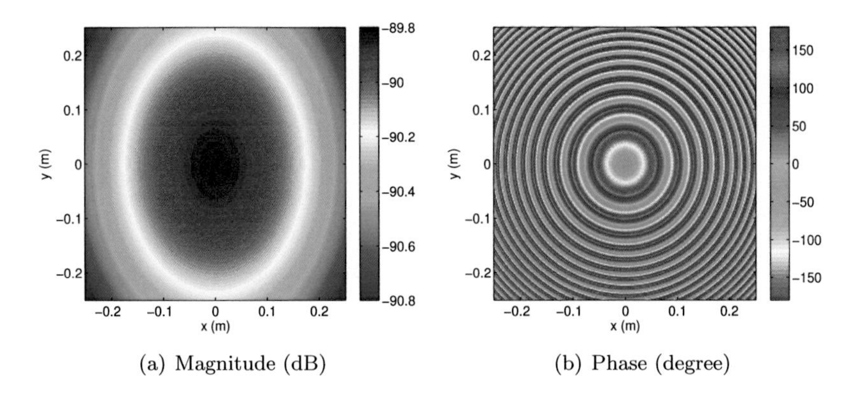


(a) Magnitude (dB) (b) Phase (degree)

Figure 4.13: Simulation results of the GO solver for the $x$-component using the same geometry shown in fig. 4.5.

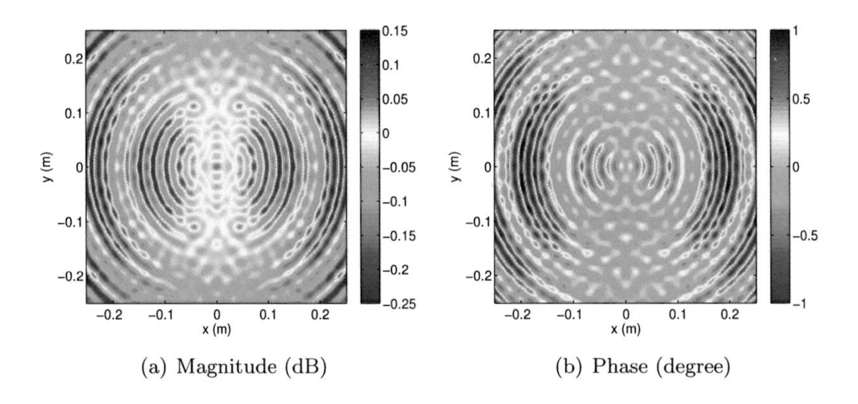


(a) Magnitude (dB) (b) Phase (degree)

Figure 4.14: Comparison between the PO and the GO solutions presented in fig. 4.6 and fig. 4.13, respectively.

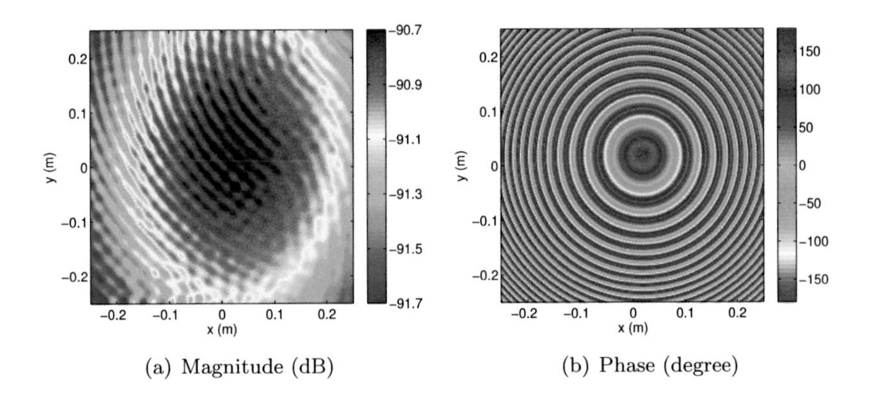


(a) Magnitude (dB) (b) Phase (degree)

Figure 4.15: PO simulation results at 80 GHz for the geometry defined in fig. 4.5, however with the Tx placed at $(0.25, 0.25, 1)$ m.

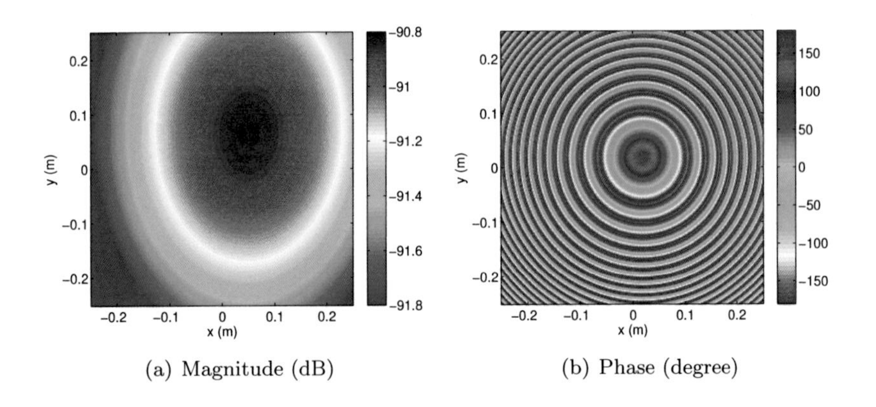


(a) Magnitude (dB) (b) Phase (degree)

Figure 4.16: GO simulation results at 80 GHz for the geometry defined in fig. 4.5, however with the Tx placed at $(0.25, 0.25, 1)$ m.

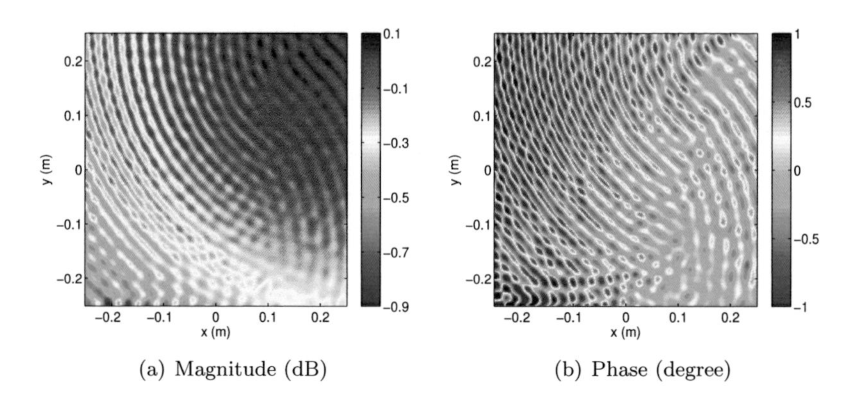


(a) Magnitude (dB)

(b) Phase (degree)

Figure 4.17: Comparison between the PO and the GO solutions presented in fig. 4.15 and fig. 4.16, respectively.

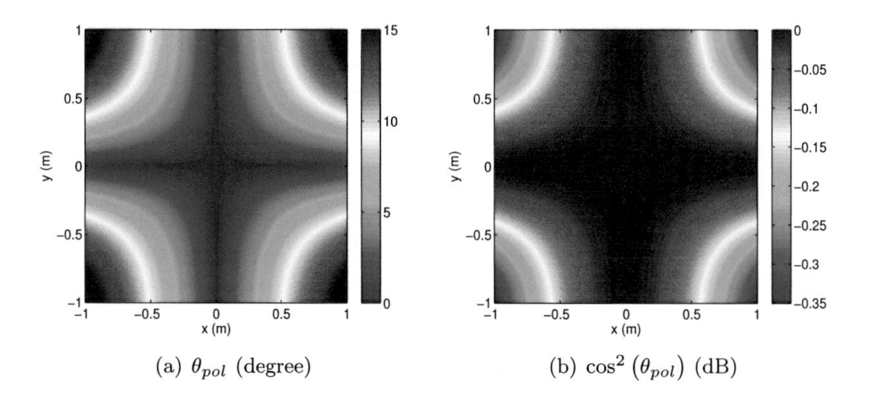


(a) $\theta_{pol}$ (degree)

(b) $\cos^2\left(\theta_{pol}\right)$ (dB)

Figure 4.18: Visualization of the polarization mismatching effect for the geometry defined in fig. 4.5.

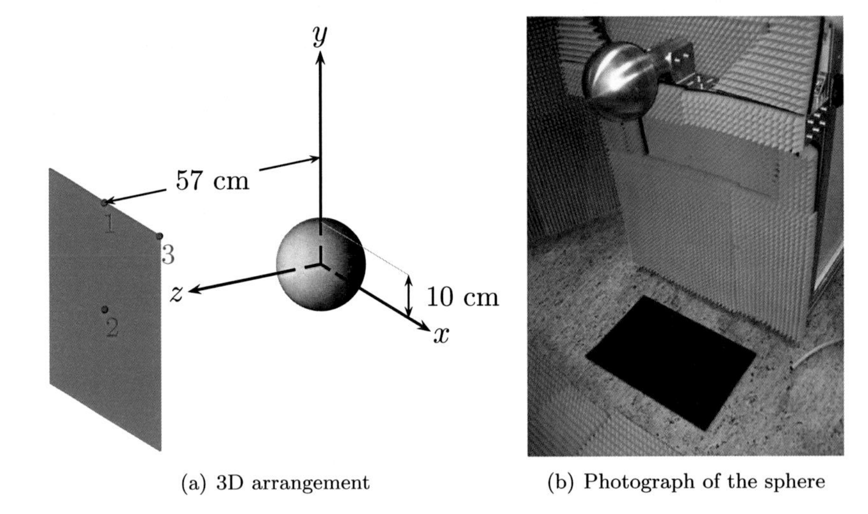


(a) 3D arrangement (b) Photograph of the sphere

Figure 4.19: Illustration of the setup used for the verification measurements, indicating the three positions of the considered transmitters and a photograph of the actual sphere with the surrounding absorbers.

The scanned aperture is 50 cm times 50 cm, and the sphere is of 10 cm radius. The sphere center is placed at 57 cm distance from the scanned aperture. During each scan, the reflection data from 75 GHz to 80 GHz is collected using two vertically polarized 13 dBi horn antennas. The reflections of the sphere are separated from the antennas' cross-coupling and the background clutter by applying a time gate to the reflection data. The gate is shaped to a Hann function in order to avoid high side-lobes after time gating. The reflections at 80 GHz are afterwards extracted from the data, and then compared with the simulations made with the GO solver while taking into account the patterns of the antennas used.

The measurement results for the upper transmitter is shown in fig. 4.20, and the corresponding simulation results is presented in fig. 4.21. A comparison between both is thus shown in fig. 4.22. Due to mechanical scanning, cable movements and positioning errors result in measurement tolerances of approximately $\pm 1$ dB in magnitude and $\pm 5^\circ$ in phase. The exact determination of the magnitude factor in the GO solution depends also on the antenna patterns, which depend in turn on the accuracy of their mechanical alignment in the scanned aperture. The comparison shows, nevertheless, a very close agreement between the measurements and the simulations. Similarly, the comparison results for the central and corner transmitters are shown in fig. 4.23 and fig. 4.24, respectively.

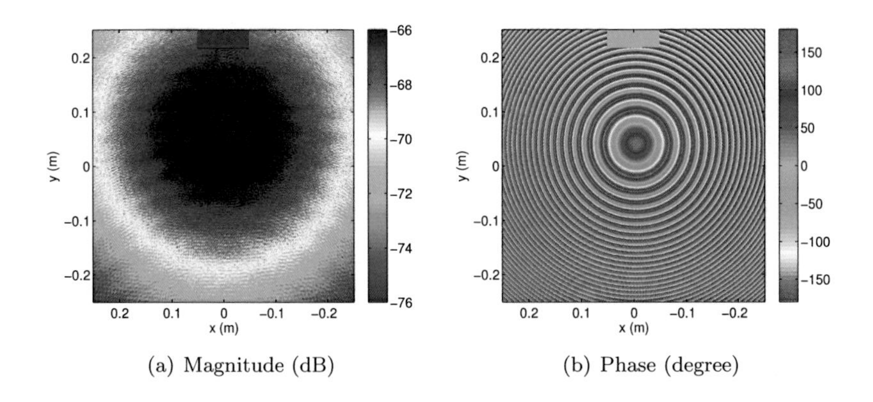


(a) Magnitude (dB) (b) Phase (degree)

Figure 4.20: Measurement results for the upper transmitter.

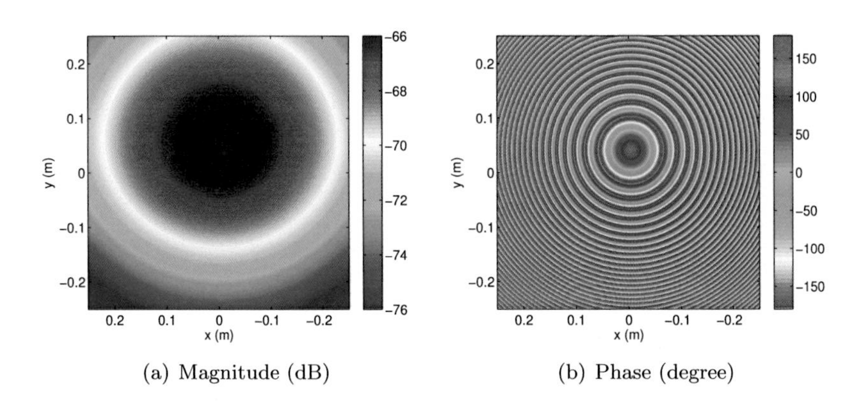


(a) Magnitude (dB) (b) Phase (degree)

Figure 4.21: Simulation results for the upper transmitter.

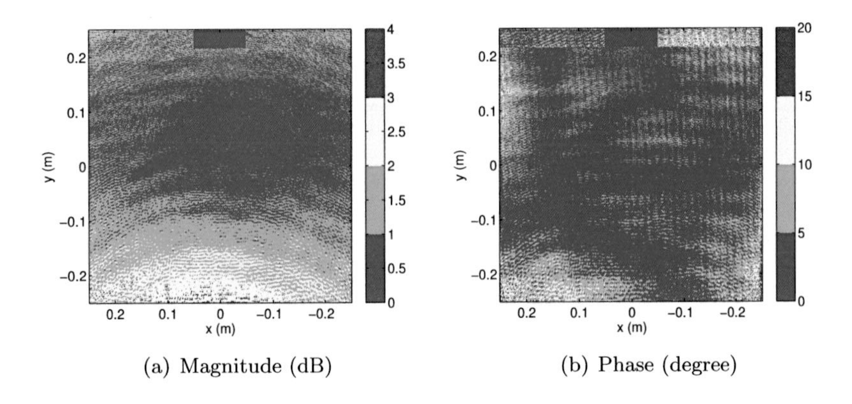


(a) Magnitude (dB) (b) Phase (degree)

Figure 4.22: Comparison results for the upper transmitter.

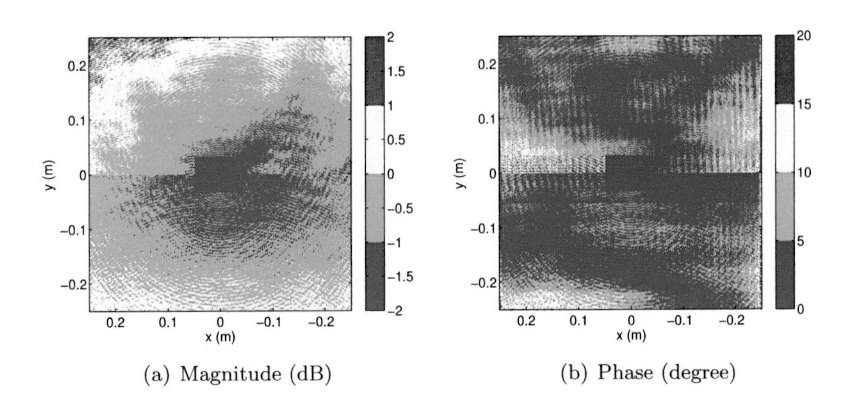


(a) Magnitude (dB) (b) Phase (degree)

Figure 4.23: Comparison results for the central transmitter.

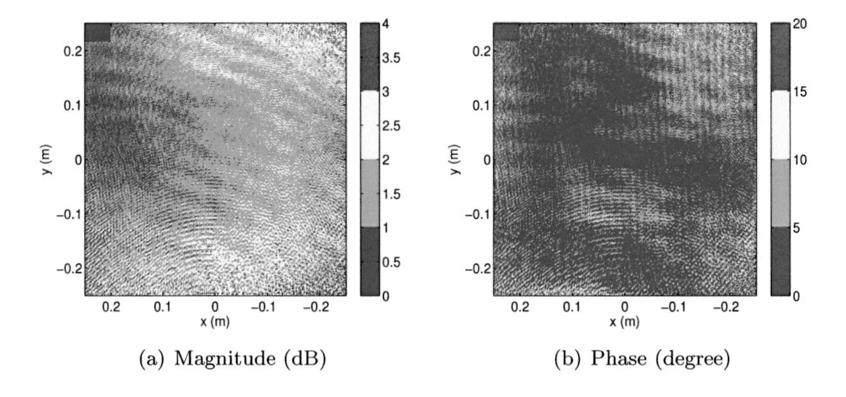


(a) Magnitude (dB) (b) Phase (degree)

Figure 4.24: Comparison results for the corner transmitter.

## 4.5 Geometry error correction

The calibration of imaging arrays is performed by using a reference reflector. The reference reflector helps, accordingly, to calculate the transfer functions for each Tx-Rx combination in the imaging array. For this, the response of the reference reflector must be precisely known, which is often reached by means of numerical simulations. In practice, the mechanical positioning of the reflector is far beyond the accuracy needed for an exact phase calibration. For example, one degree of phase at 75 GHz corresponds to $5.6\times10^{-6}$ m of mechanical displacement. The reflector can be a plate, where the assumption of exact parallel positioning to the array aperture is also influenced by alignment errors. In spite of the high requirements on the phase measurement accuracy in synthetic aperture processing, the phase errors made by the mechanical positioning are tolerable. The reason behind this lies in the fact that the reflector by itself delivers a reference wavefront as a whole, therefore any displacement from its ideal position counts to a systematic phase error in the whole array. This leads to a calibration based on a wavefront, which is assuming another curvature than the actual wavefront is. Depending on the geometry of the reflector and the distance to the array, the resultant degree of mismatch can be found and analyzed. It is even possible to go further than this and try to correct the calibration data, i.e., to increase the knowledge about the position of the reflector used. This leads to an unusual situation where the task is to find the actual position of the reflector using an imaging system which is not yet calibrated; and afterwards use this information to calibrate the system itself. This is only mathematically possible if some sort of an error function can be constructed from the calibration data, and hence to be used to evaluate the calibration quality itself. In order to reach a sort of a representative error function, the imaging system operation must be guaranteed even if the calibration is not yet optimal. The error function is also better chosen to represent directly the imaging quality of the system. For this, it must be proved in the first place, that the imaging system is capable to operate under the condition of being calibrated using a non-optimally positioned reflector. Addressing the sphere as a specially suitable reflector for the imaging systems under consideration, the analysis will be targeting it, however the results also apply to other calibration objects. Flat plates can be considered as a special case of a sphere with an infinite radius.

In the following, the discussion will go through the effects associated with the displacement of the calibration sphere from its ideal position, and the associated influences on the image quality. Afterwards, a suggestion for error functions will be made and verified numerically. Upon this, a correction algorithm for the geometrical errors is presented.

### 4.5.1 Mathematical development

The calibration of the array aperture is based on illuminating a sphere and comparing the measured reflections with the simulation of the sphere reflections. Having the sphere in a slightly different position than the one considered in the simulation, introduces a mismatch in magnitude and phase, which accordingly deforms the reference plane. The mismatch in magnitude is negligible in this case. The phase can, however, change with several $2\pi$ depending on the displacement level and direction. Phase mismatch gets larger for displacements in the range direction than the cross-range one. Nevertheless, the phase mismatch is systematic across the array aperture. This systematic phase mismatch can be mathematically traced, and hence its influences on the image reconstruction process can be analyzed. Hereafter, for the sake of simplicity, the positioning error is assumed in $xz$-plane, thus including the direction along range as well as cross-range. The results can then be easily generalized for the possible three-dimensional positioning errors.

Fig. 4.25 illustrates the calibration procedure of a single transmitter. The transmitter illuminates the calibration sphere with a spherical wavefront, which in turn reflects back to the array. The reflected wavefront in this case is strictly not spherical, as discussed in sec. 4.4. In the paraxial regime, the wavefront is almost spherical and the center of the reflected wave lies inside the sphere between its focal point and its surface, as depicted in sec. 4.4.2.4. However, this is used as an approximation of the wavefront in order to achieve a closed form description.

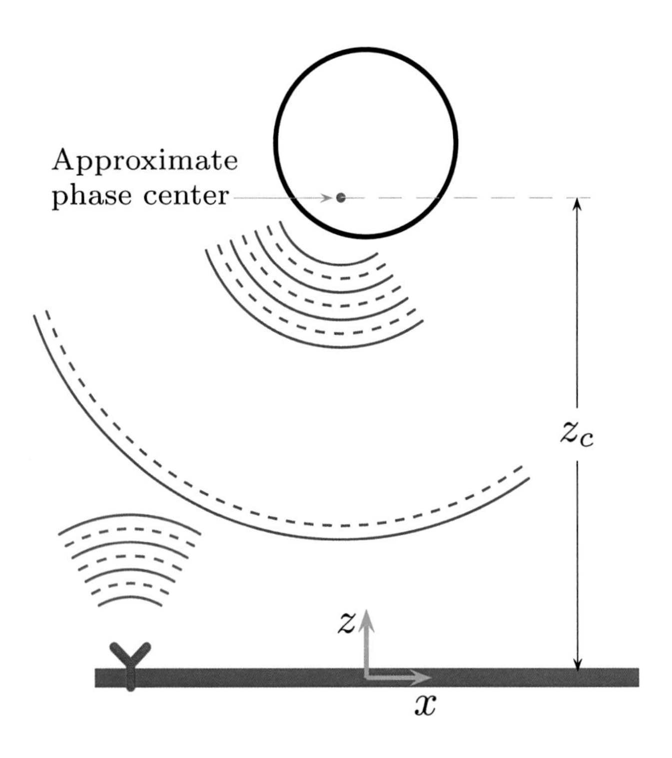


Figure 4.25: Example of the wavefronts used during the calibration procedure. As a simplification, an approximate phase center can be assumed to lay inside the sphere volume.

The distance between the center of the reflected wave back to the array and the phase centers of the antennas is named $z_c$, as annotated in fig. 4.25. The sphere is tended to be placed centrally to the array aperture. Due to the mechanical positioning, small displacements of $\Delta x_c$ and $\Delta z_c$ are expected, thus along cross-range and range directions, respectively. The error in the calculation of the wavefront of the reflected wave includes a constant phase difference due to the change in the travel time for the transmitted wave, and a spatial varying part due to the displacement of the center of the reflected spherical wavefront. Considering the varying part for a wavefront with the phase center $(\Delta x_c, 0, z_c + \Delta z_c)$, the error is described using the difference $\Delta d$ in the geometrical lengths between the cases of ideal and actual positioning, readily as

$$\Delta d = \underbrace{\sqrt{(x + \Delta x_c)^2 + y^2 + (z_c + \Delta z_c)^2}}_{\text{actual wavefront}} - \underbrace{\sqrt{x^2 + y^2 + z_c^2}}_{\text{ideal wavefront}} . \tag{4.38}$$

For small $\Delta x_c$ and $\Delta z_c = 0$, the geometrical difference can be approximately rewritten as

$$\begin{aligned} \Delta d(\Delta x_c, \Delta z_c = 0) &\approx \Delta x_c \cdot \frac{\partial}{\partial x}(\sqrt{x^2 + y^2 + z_c^2}) \\ &\approx \Delta x_c \cdot \frac{x}{\sqrt{x^2 + y^2 + z_c^2}} . \end{aligned} \tag{4.39}$$

Similarly, for small $\Delta z_c$ and $\Delta x_c = 0$, it yields

$$\begin{aligned} \Delta d(\Delta x_c = 0, \Delta z_c) &\approx \Delta z_c \cdot \frac{\partial}{\partial z_c}(\sqrt{x^2 + y^2 + z_c^2}) \\ &\approx \Delta z_c \cdot \frac{z_c}{\sqrt{x^2 + y^2 + z_c^2}} . \end{aligned} \tag{4.40}$$

Considering $z_c \gg \sqrt{x^2 + y^2}$, the error descriptions simplify[§] to

$$\Delta d(\Delta x_c, \Delta z_c = 0) = \Delta x_c \cdot \left( \frac{x}{z_c} \right) \quad \text{and} \tag{4.41}$$

$$\Delta d(\Delta x_c = 0, \Delta z_c) = \Delta z_c \ . \tag{4.42}$$

The error expressed in eq. 4.41 results in a linear phase shift along the aperture and over frequency. This linear phase shift causes less spatial

[§]This step is required to linearize the error descriptions and hence to emphasize the dominant factors.

shift in the reconstructed image, which shifts in the opposite direction to the displaced calibration sphere. This is associated with some defocusing due to the phase shift over frequency. Therefore, the image will partially follow the displacement of the calibration sphere with some reduction in quality. Though, the error expressed in eq. 4.42 causes a constant phase shift along the aperture, however a linear one over frequency. This results in an equivalent displacement of the reference plane with $\Delta z_c$ along range direction. Consequently, the reconstructed image will be equally shifted along range direction, however without being exposed to defocusing effects. In summary, for small displacements of the calibration object, the phase mismatches cause defocusing effects to the reconstructed images, however do not hinder the image production. This indicates the robustness of the calibration procedure despite mechanical misalignment. The exact loss in the focusing quality depends on the imaging parameters, and is examined numerically in the next section.

On the contrary to the linear portion of the error, nonlinear error at the outer part of the aperture is there where $\sqrt{x^2 + y^2}$ is comparable to $z_c$, hence eq. 4.39 and eq. 4.40 apply instead. These parts of the aperture will not be fully coherently processed. Defocusing, blurring, and loss of image quality are expected due to the nonlinear deformation of the reference phase plane. Nevertheless, image production is still ensured by the inner part of the aperture, where linear effects are dominant.

### 4.5.2 Numerical verifications

The influence of the deformation in the reference phase plane on the image quality is now examined numerically. For this, the second array introduced in sec. 2.8.4 is considered. The numerical verifications are made for a defined calibration sphere geometry, i.e., radius and position. A mechanical displacement is assumed, and hence the response of the sphere is simulated twice, i.e., for the actual position and the ideal one. The error function $\xi$ is calculated as the ratio between the two simulations, as

$$\xi = \frac{\Gamma_{\text{sim}}^{\text{ideal}}}{\Gamma_{\text{sim}}^{\text{actual}}} \ . \tag{4.43}$$

In order to examine the imaging process, a simulation is performed for a grid of isotropic reflectors placed at a distance of 0.5 m in front of the array. This is illustrated in fig. 4.26, where 11×11 reflectors in equidistant raster are used. According to the calibration procedure in sec. 4.2, the simulated data $s_{\text{sim}}^{\text{ideal}}$ is modified as

$$s_{\text{sim}}^{\text{actual}} = \xi \cdot s_{\text{sim}}^{\text{ideal}} \ . \tag{4.44}$$

Afterwards, the $s_{\text{sim}}^{\text{actual}}$ goes through the reconstruction process.

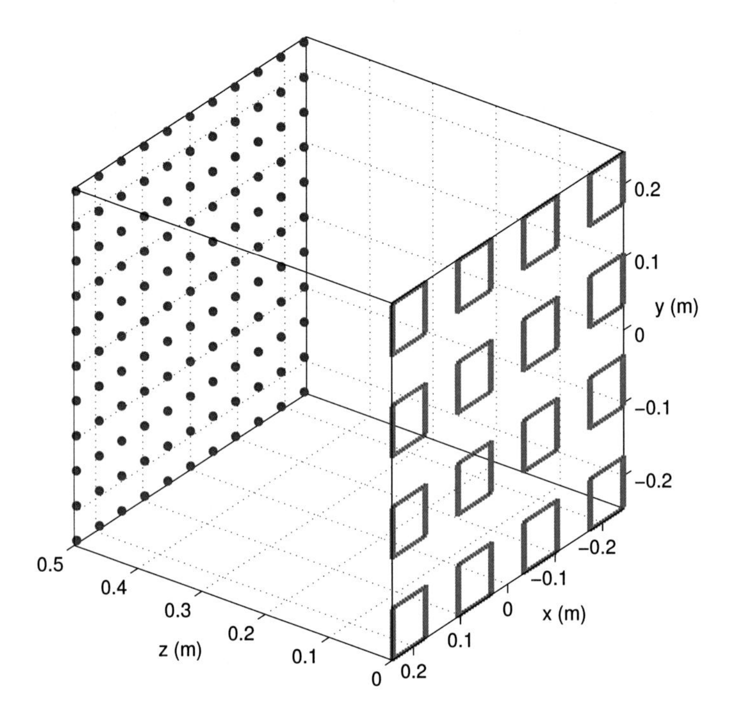


Figure 4.26: The geometrical arrangement used in the numerical verifications for the calibration sphere displacement.

The simulation is performed from 75 to 82 GHz using 32 frequency steps. The calibration sphere has a radius of 100 mm with its center positioned centrally on a distance of 600 mm away from the array. Fig. 4.27(a) shows the reconstructed image for the ideal case when no mechanical displacement takes place. The 11×11 reflectors are clearly reconstructed and viewed in a 40 dB image. In fig. 4.27(b), the sphere center is assumed to be displaced by 4 mm and 2 mm in $x$- and $z$-directions, respectively. However the displacement is comparable to the wavelength, the image reconstruction is successful. The reflectors image is partially shifted following the displacement in the calibration sphere, as detailed in fig. 4.28 for the central reflector.

Considering a case with larger positioning errors, fig. 4.29(a) shows the resultant image for displacement of 20 mm in $x$-, 40 mm in $y$-, and 10 mm in $z$-directions. The defocusing effects are clearly seen, specially away from the image center. Loss in resolution and intensity are there. The image is also globally shifted. Fig. 4.29(b) illustrates the image of the shifted central reflector.

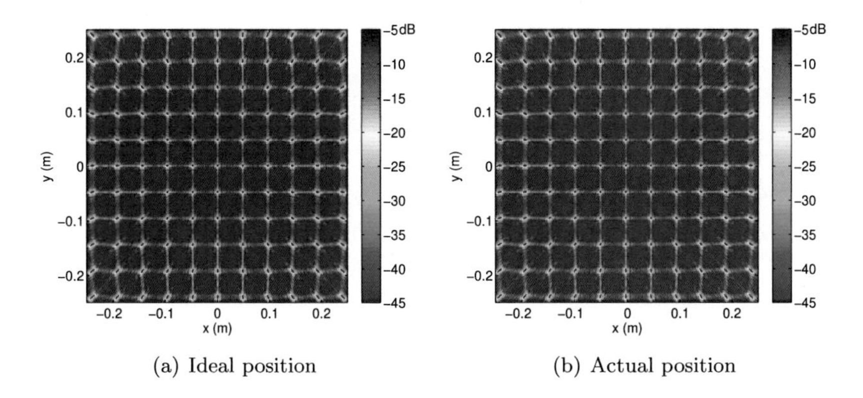


(a) Ideal position

(b) Actual position

Figure 4.27: On the left side, simulation result of 11×11 reflectors assuming ideal position of a calibration sphere. On the right side, the center of the sphere is displaced by 4 mm and 2 mm in $x$- and $z$-directions, respectively.

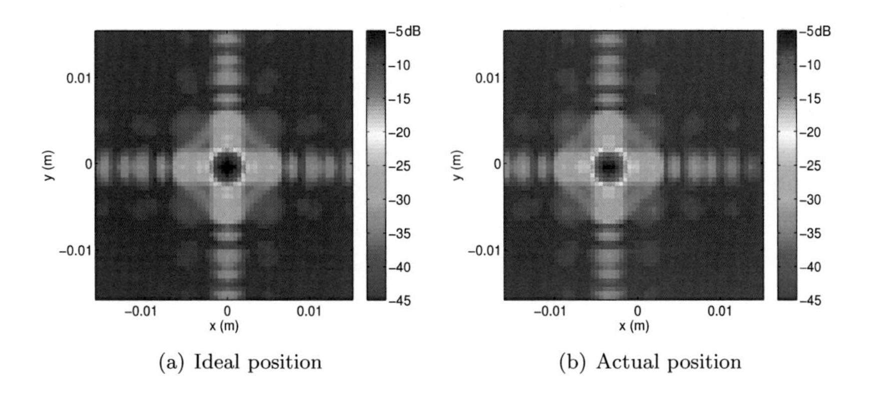


(a) Ideal position

(b) Actual position

Figure 4.28: Detailed view of the central reflector presented in fig. 4.27. The image is slightly shifted following the displaced calibration sphere, however the good image quality is still maintained.

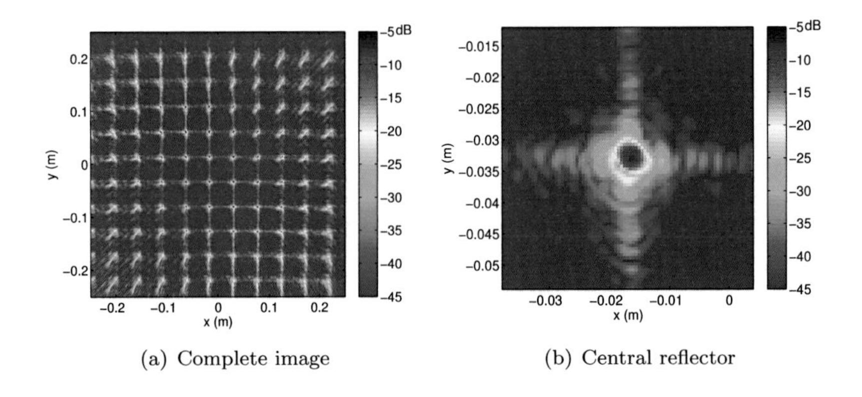


(a) Complete image

(b) Central reflector

Figure 4.29: Simulation result using a calibration sphere with an extreme displacement of 20 mm in $x$-, 40 mm in $y$-, and 10 mm in $z$-directions.

### 4.5.3 Geometry search algorithm

The loss in the image quality, discussed above, can be reduced or even fully avoided if the position of the calibration object, e.g., the sphere, could become known. In this case, the simulation used during the calibration procedure will use the actual position of the sphere and no deformation in the reference phase plane will further occur. This is specifically critical for the imaged objects which are placed far from the array center. As seen in the numerical simulations, it is sufficient to position the sphere with an accuracy close to the used wavelength, e.g., a few millimeters, to ensure image focusing afterward. In order to avoid the usage of sophisticated mechanical alignment, numerical solutions are better established to find the actual position of the sphere and hence correct the calibration geometry.

The following solution is proposed to search for the actual position of the sphere center. Small reflectors will be added to the scene to reveal the image distortion. These reflectors are used for the generation of a test image, which delivers a feedback about the image quality. They must be accurately positioned relative to the calibration object, however not necessarily accurately positioned relative to the array. While analyzing the image distortion happening to the test reflectors, knowledge about the actual displacement of the calibration object is achieved. For instance, small reflectors can be placed around the calibration sphere, or hemisphere, as illustrated in fig. 4.30. The size of the reflectors should be comparable to the used wavelength, in order to ensure diffuse reflections and hence a good visibility from all aperture parts. The visibility of the small reflectors is guaranteed by the different levels of processing gains discussed in chapter 3. The calibration measurement of the sphere is thus followed by a second measurement where the reflectors are attached around. Some initial displacement vector is then assumed, e.g., null vector in the first step. The reconstructed image can then be evaluated and a measure representing the quality of the image is thus computed. Upon this, an iterative search[§] for the actual position of the sphere center can take place. This is illustrated by the flow chart in fig. 4.34. For the success of the iterative search, a representative evaluation function must be defined. This function can be constructed using one or a combination of:

- The intensities of the imaged reflectors.
- The geometrical exactness for the relative positioning of the reflectors. Any stretching or compression in the image can be evaluated.
- The sharpness of the imaged reflectors in comparison to the expectation for the image resolution.

In order to illustrate these effects, a numerical simulation is shown next, in which a hemisphere of 50 mm radius is surrounded with 12 small spherical reflectors each of 2 mm radius. The small reflectors are positioned angularly equispaced on a concentric circle of 200 mm radius with the hemisphere. The plane including the reflectors is parallel to the array aperture, as shown in fig. 4.31. The array geometry and the used frequencies are the same as in sec. 4.5.2. Fig. 4.32(a) shows the reconstructed image when the calibration object is assumed to be ideally positioned, i.e., with null displacement. Applying the procedure explained above, the reconstructed image, assuming a displacement of the calibration object equal to 20 mm, 40 mm, and 10 mm in $x$-, $y$-, and $z$-directions, respectively, is shown in fig. 4.32(b). By comparing both images, the image of the reflectors keeps centered around the calibration object for the reasons explained above. It comprises, however, some distortion. A detailed view is shown in fig. 4.33, in which the loss in resolution and intensity can be clearly observed. These are examples for the image features to be used in the construction of an evaluation function serving for the feedback mechanism.

It is often possible to consider the image distortions caused by the displacement in the three dimensions to be separable. Being a matter of approximation, the results in sec. 4.5.1 suggest this separability when the linear effects are dominant.

[§]This iterative search is conceptually similar to the automatic focusing techniques used in optical domain, in which one or more parameters in the imaging system are modified in order to enhance the image quality in terms of resolution and/or dynamic range.

Figure 4.30: Illustration of a calibration hemisphere surrounded with small reflectors used to assist the geometry search algorithm.

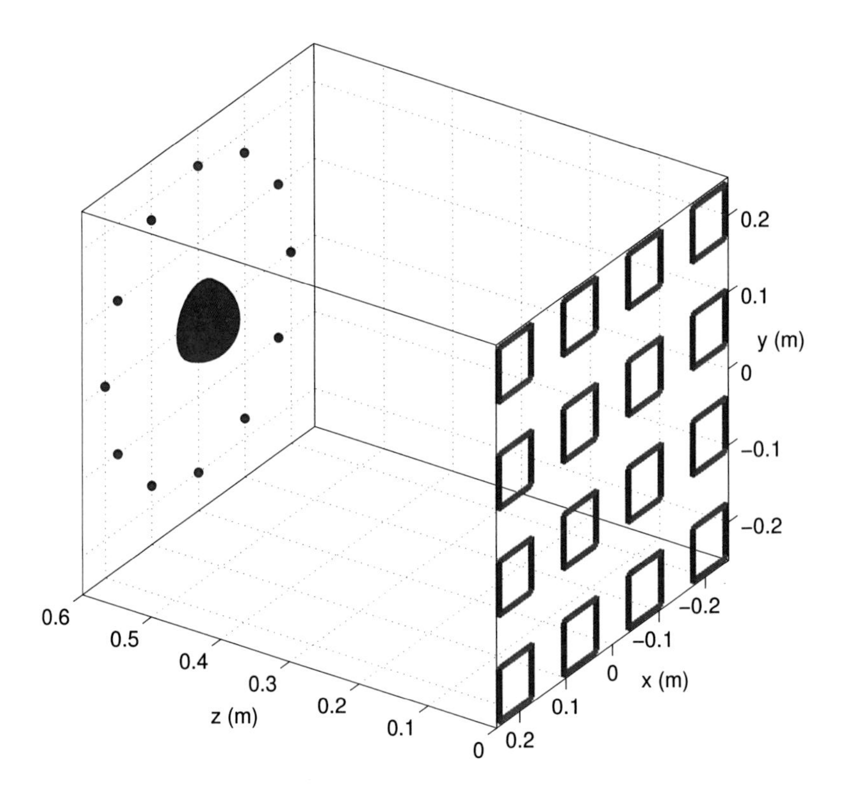


Figure 4.31: The geometrical arrangement for the simulation of the calibration hemisphere, while being surrounded by 12 small reflectors.

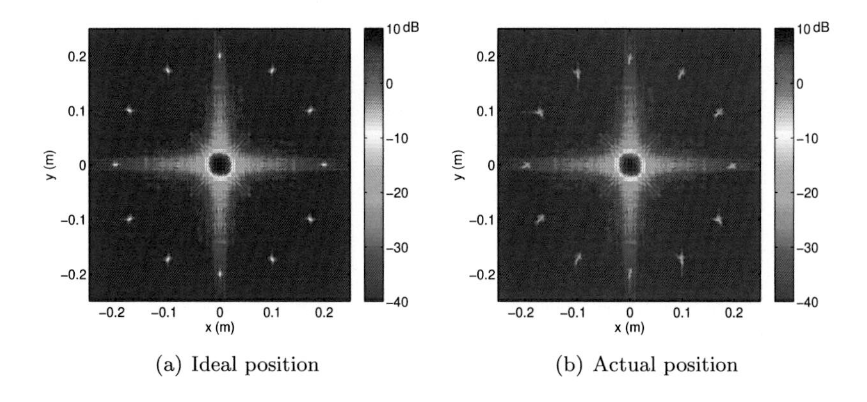


Figure 4.32: On the left side, simulation result while the calibration hemisphere is ideally positioned. On the right side, the image of the objects is shown after assuming a displacement of 20 mm in $x$-, 40 mm in $y$-, and 10 mm in $z$-directions.

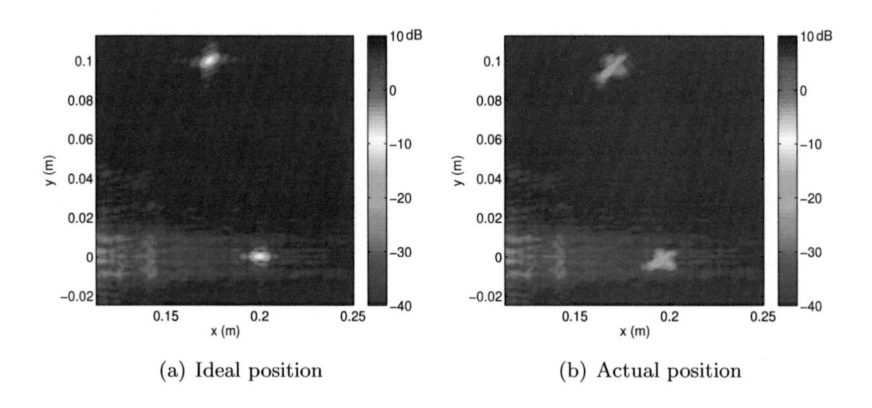


Figure 4.33: Detailed view for the upper right part of the reflectors simulated in fig. 4.32.

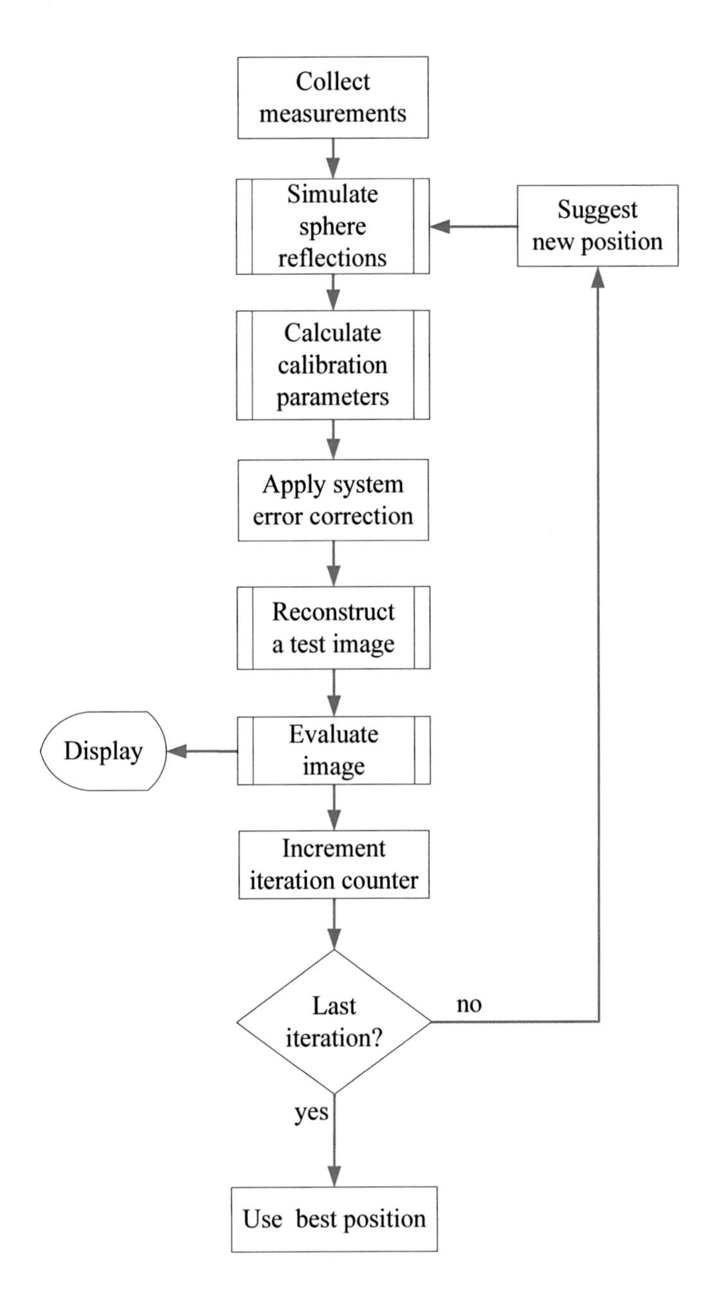


Figure 4.34: Flow chart illustrating the iterative search of the sphere position.

# 4.6 Calibration matrix refinement

During the calibration of multistatic arrays, the calibration matrix $\mathbf{T}$ is calculated. The matrix $\mathbf{T}$ represents the transfer function of each Tx-Rx combination in the imaging array. For $N_t$ transmitters, and $N_r$ receivers the matrix $\mathbf{T}$ has $N_t \cdot N_r$ elements at each frequency. The imaging system has a total number of $N_t + N_r$ channels and therefore $N_t + N_r$ independent unknowns. For large arrays, this leads to redundancy in the collected information regarding the characteristics of the channels. The redundant information is as large as $N_t \cdot N_r - (N_t + N_r)$. The redundancy can be used to enhance the SNR of the calculated transfer functions. An estimate $\widehat{\mathbf{T}}$ can be achieved by means of averaging the available redundant information at each frequency. Therefore, a noisy calibration matrix can be refined mathematically, where the refinement process is to be considered at each frequency separately. In the following analysis, the $\widetilde{\cdot}$ symbol indicates noisy data, e.g., includes thermal noise, and $\widehat{\cdot}$ symbol indicates a statistical estimate. For the sake of simplicity, it will be considered that $N_t = N_r = N$, which leads to a squared $\mathbf{T}$ matrix. The method can be easily generalized for non-equal numbers of transmitters and receivers.

## 4.6.1 Mathematical development

The complex value of $\widetilde{T}$ is calculated from

$$\widetilde{T} = \frac{\widetilde{M_s} - \widetilde{M_m}}{\Gamma_{\text{sim}}}. \tag{4.45}$$

The noisy measurements $\widetilde{M_s}$ and $\widetilde{M_m}$ performed during the calibration include additive white Gaussian noise (AWGN). On the contrary, the $\Gamma_{\text{sim}}$ is free from thermal noise as it is produced by simulation. In the denominator of eq. 4.45, the values of $\Gamma_{\text{sim}}$ has almost constant magnitudes and strong varying phases. Because the phase comes from a plane cut in the far-field of a spherical wave over several order of magnitudes of wavelengths, the phase factor is almost equally distributed from $-\pi$ to $+\pi$. Therefore, the probability density function (PDF) of the Gaussian noise will only be scaled with $|\Gamma_{\text{sim}}|$ without undergoing shape changing. In the nominator of eq. 4.45, the noise in $\widetilde{T}$ is accumulated through linear operation on two independent AWGN signals and therefore the resultant noise is also AWGN. Let the standard deviation of the thermal noise per a single measurement is equal to $\sigma_{\text{M}}$, the noisy $\widetilde{T}$ with corresponding AWGN function $\mathcal{N}$(mean, variance) is hence described as

$$\widetilde{T} = T + \mathcal{N}\left(0\,,\ 2\left(\frac{\sigma_{\text{M}}}{|\Gamma_{\text{sim}}|}\right)^2\right). \tag{4.46}$$

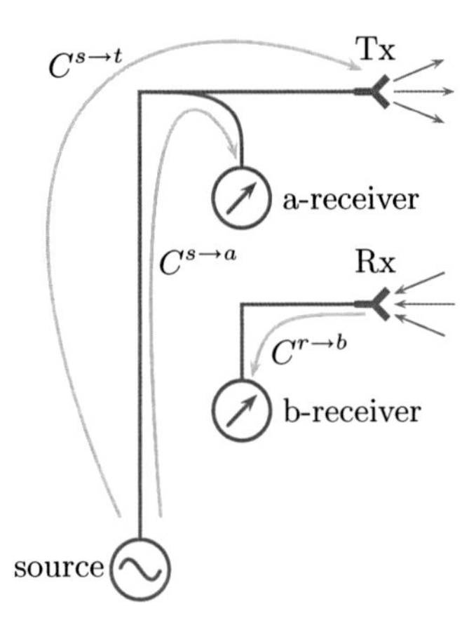


Figure 4.35: Decomposition of $T$ transfer function.

The transfer function $T$ can be decomposed in two characteristic transfer functions, namely $C^t$ for transmitters and $C^r$ for receivers. The signals flow is indicated in fig. 4.35, and the result can be expressed as

$$T = \frac{b}{a}|_{\Gamma_{\mathrm{DUT}=1}} = \frac{C^{s\to t} \cdot 1 \cdot C^{r\to b}}{C^{s\to a}} = \underbrace{\frac{C^{s\to t}}{C^{s\to a}}}_{C^t} \cdot \underbrace{C^{r\to b}}_{C^r}. \tag{4.47}$$

On one hand, each Tx number $n$ has a characteristic transfer function $C_n^t$ describing the output signal at the corresponding Tx antenna relative to the source and normalized to the corresponding reference channel. On the other hand, each Rx number $m$ has a characteristic transfer function $C_m^r$ describing the measured signal at the receiver relative to the input signal from the corresponding Rx antenna. The matrix $\mathbf{T}$ is hence described as

$$\begin{aligned}\mathbf{T}|_{N\times N} &= \mathbf{C}^t|_{N\times 1} \cdot \mathbf{C}^r|_{1\times N} \\ &= \begin{bmatrix} C_1^t C_1^r & C_1^t C_2^r & \cdots & C_1^t C_N^r \\ C_2^t C_1^r & C_2^t C_2^r & \cdots & C_2^t C_N^r \\ C_3^t C_1^r & C_3^t C_2^r & \cdots & C_3^t C_N^r \\ \vdots & \vdots & \ddots & \vdots \\ C_N^t C_1^r & C_N^t C_2^r & \cdots & C_N^t C_N^r \end{bmatrix},\end{aligned} \tag{4.48}$$

where

$$\mathbf{C}^t|_{N\times 1} = \begin{bmatrix} C_1^t \\ C_2^t \\ \vdots \\ C_N^t \end{bmatrix} \quad \text{and} \quad \mathbf{C}^r|_{1\times N} = [C_1^r C_2^r \dots C_N^r] \,. \tag{4.49}$$

During the calibration, direct access to $C_n^t C_m^r$ values is not feasible, and instead they are available with AWGN, i.e., as $\widetilde{C_n^t C_m^r}$.

In order to get an estimate $\widehat{C_m^r}$, one can consider

$$\widehat{\left(\frac{C_m^r}{C_k^r}\right)} = \frac{1}{N}\sum_{n=1}^{N}\frac{\widetilde{C_n^t C_m^r}}{\widetilde{C_n^t C_k^r}} \,, \tag{4.50}$$

where $k$ can be arbitrarily chosen from 1 to $N$. Here, the estimate is achieved by averaging the available records in the column $m$ of $\widetilde{\mathbf{T}}$ relative to a reference column $k$. Similarly, an estimate for $\widehat{C_n^t}$ can be found using

$$\widehat{\left(\frac{C_n^t}{C_k^t}\right)} = \frac{1}{N}\sum_{m=1}^{N}\frac{\widetilde{C_n^t C_m^r}}{\widetilde{C_k^t C_m^r}} \,, \tag{4.51}$$

where $k$ is also arbitrarily chosen. In this case, the estimate is found by averaging the row $n$ relative to another row $k$.

It is preferable to formulate the refinement process in a matrix form. For readability, the operator $\Lambda$ is defined as

$$\mathbf{Y} = \Lambda\left(\mathbf{X}\right) \,, \tag{4.52}$$

where $\mathbf{Y}$ is the element-wise inversion matrix of a transposed matrix $\mathbf{X}$.

Knowing initial estimates $\widehat{\mathbf{C}^t}^0$ and $\widehat{\mathbf{C}^r}^0$, the next ones $\widehat{\mathbf{C}^t}^1$ and $\widehat{\mathbf{C}^r}^1$ can be calculated. Following eq. 4.49, the estimates are thus related by

$$\widehat{\mathbf{T}}^0 = \widehat{\mathbf{C}^t}^0 \cdot \widehat{\mathbf{C}^r}^0 \quad \text{and} \tag{4.53a}$$

$$\widehat{\mathbf{T}}^1 = \widehat{\mathbf{C}^t}^1 \cdot \widehat{\mathbf{C}^r}^1 \,. \tag{4.53b}$$

Using a matrix formulation based on eq. 4.50 and eq. 4.51, the next estimates are related to the previous ones by

$$\widehat{\mathbf{C}^t}^1 = \frac{1}{N} \cdot \widehat{\mathbf{T}}^0 \cdot \Lambda\left(\widehat{\mathbf{C}^r}^0\right) \quad \text{and} \tag{4.54a}$$

$$\widehat{\mathbf{C}^r}^1 = \frac{1}{N} \cdot \Lambda\left(\widehat{\mathbf{C}^t}^0\right) \cdot \widehat{\mathbf{T}}^0 \,. \tag{4.54b}$$

Applying eq. 4.54 in eq. 4.53 and after some algebraic arrangements, the estimates of $\widehat{\mathbf{T}}$ are related according to

$$\widehat{\mathbf{T}}^1 = \frac{1}{N^2} \cdot \widehat{\mathbf{T}}^0 \cdot \Lambda\left(\widehat{\mathbf{T}}^0\right) \cdot \widehat{\mathbf{T}}^0 . \tag{4.55}$$

The next estimate of the calibration matrix is hence directly calculated from the previous one by double matrix multiplication followed by a normalization factor equal to the total number of elements inside the matrix. In this formulation, it is no more necessary to decompose $\widetilde{\mathbf{T}}$ into the characteristic functions $\mathbf{C}^t$ and $\mathbf{C}^r$, and the result is therefore independent of any arbitrary choice of a reference column or row as in eq. 4.50 and eq. 4.51. Using the calibration matrix $\widetilde{\mathbf{T}}$, according to eq. 4.46, as an initial estimate, then the enhanced estimate $\widehat{\mathbf{T}}$ is given by

$$\widehat{\mathbf{T}} = \frac{1}{N^2} \cdot \widetilde{\mathbf{T}} \cdot \Lambda\left(\widetilde{\mathbf{T}}\right) \cdot \widetilde{\mathbf{T}} . \tag{4.56}$$

The noise in the data is assumed to be fully uncorrelated, thus the limit reduction of its variance is given by $\frac{N^2}{2N} = N/2$; because $N^2$ records are averaged to get $2N$ ones. Following the definition in eq. 4.46, the noise in the records of the estimate of the transfer function after the refinement process is expressed by

$$\widehat{T} = T + \mathcal{N}\left(0\,,\ \frac{4}{N}\left(\frac{\sigma_{\mathrm{M}}}{|\Gamma_{\mathrm{sim}}|}\right)^2\right) . \tag{4.57}$$

### 4.6.2 Practical considerations

The formulation in eq. 4.55 offers an iterative solution for the refinement process. However this is generally true, the iterative process would not help much to enhance the estimate after the first iteration. The reason for this lies in the fact that the refinement process assumes a zero mean uncorrelated noise, which is strictly true only for $\widetilde{\mathbf{T}}$. The process in eq. 4.55, or similarly in eq. 4.50 and eq. 4.51, performs an averaging operation in order to refine the records. This averaging is made to noisy records with AWGN. As a result, the estimated record will not include AWGN, on the contrary a biased noise, e.g., with non-zero mean. This biasing effect, on one hand reduces the performance of further iterations, and on the other hand constrains the requirement of a minimum SNR level in $\widetilde{\mathbf{T}}$ for the success of the refinement process. In other words, the initial values of the $\widetilde{C^t_n C^r_m}$ should not be dominated by noise.

The enhancement of the SNR described in eq. 4.57 is very valuable for the calibration of large arrays. Imaging arrays often include hundreds to thousands of elements. For an array with 1000 Tx and 1000 Rx, the SNR could be enhanced with a factor of $10\log_{10}(1000/2) \approx 27$ dB. This factor can be used to relax the requirements on the integration time in the receivers, and hence speed up the calibration procedure, or allow for the usage of smaller calibration objects with lower RCS values, or alternatively increase the distance to the calibration object. Depending on the calibration scenario, a combination of these choices can also be decided.

In practice, the calibration matrix $\widetilde{\mathbf{T}}$ could be sparse either due to the existence of defect channels, or invalid data made by receiver compression at places of strong cross-coupling with neighboring transmitters. These records are usually replaced with nulls in order to avoid their influence on the image reconstruction process. Direct application of eq. 4.56 will then undergo instability and result in divide by zero errors, i.e., singularities in $\Lambda(\widetilde{\mathbf{T}})$. Therefore, the numerical implementation of the refinement process must take into account the sparseness of the matrix and also correct for the normalization factor where some records are physically not available for the averaging process. This can be achieved by applying the following steps:

1. Replacing invalid records in $\widetilde{\mathbf{T}}$ with nulls.

2. The corresponding records in the result of $\Lambda(\widetilde{\mathbf{T}})$ due to the nulls in $\widetilde{\mathbf{T}}$ must be replaces with nulls again before being applied to the matrix multiplication step.

3. Instead of using the single value $N^2$, a normalization matrix $\mathbf{N}$ is to be used. The normalization is an element-wise operation. The matrix $\mathbf{N}$ can be easily found by considering the case when all valid records are ideally equal to unity, in this case the matrix multiplication used in the refinement process will yield the normalization matrix itself. The inversion made by $\Lambda$ must be avoided, however. This is described mathematically by

$$\mathbf{N} = \left(\widetilde{\mathbf{T}} \neq 0\right) \cdot \left(\widetilde{\mathbf{T}} \neq 0\right)^{\mathrm{tr}} \cdot \left(\widetilde{\mathbf{T}} \neq 0\right), \tag{4.58}$$

where $\left(\widetilde{\mathbf{T}} \neq 0\right)$ indicates the logical operation resulting in ones when a valid record is available in the matrix $\widetilde{\mathbf{T}}$ and zero otherwise, and $()^{\mathrm{tr}}$ indicates matrix transpose operation.

### 4.6.3 Numerical verification

The performance of the calibration matrix refinement algorithm can be investigated by means of numerical simulations. For this, a synthetic column vector $\mathbf{C}^t$ and a row vector $\mathbf{C}^r$ are generated with little variations in the magnitude and strong variations in the phases, which simulates the real situation in an imaging system at some frequency of operation. However these values are deterministic in nature, the task of the algorithm is to come to an estimate as close as possible of their product. The noiseless calibration matrix $\mathbf{T}$ and the noisy matrix $\widetilde{\mathbf{T}}$, with added complex AWGN, are simulated by

$$\mathbf{T} = \mathbf{C}^t \cdot \mathbf{C}^r \tag{4.59a}$$

$$\widetilde{\mathbf{T}} = \mathbf{T} + \mathcal{N}(0,1) \; . \tag{4.59b}$$

In order to examine the performance of the algorithm, the noise content of the simulated $\widetilde{\mathbf{T}}$ and the estimated $\widehat{\mathbf{T}}$ are to be compared. The rest noise in $\widehat{\mathbf{T}}$ is calculated from $\widehat{\mathbf{T}} - \mathbf{T}$. The simulation is hence made using the following simulation parameters:

- 768 transmitters and 768 receivers
- values of $|C^t|$ and $|C^r|$ are uniformly distributed, and centered at $\sqrt{10}$ with $\pm 10\%$
- phases of $\angle C^t$ and $\angle C^r$ are uniformly distributed from $-\pi$ to $+\pi$

Performing the refinement process as in eq. 4.56, the real part of the noise before and after the refinement process is shown in fig. 4.36(a), the result is similar for the imaginary part. The figure shows the strong reduction in the variance of the PDF of the noise, without visible change in the zero-mean Gaussian shape of the distribution. The variance of the rest noise in $\widehat{\mathbf{T}}$ is reduced to 0.0026, instead of the unity variance in $\widetilde{\mathbf{T}}$, which is equivalent to a reduction in the noise power by 25.9 dB. This agrees with the expected theoretical limit reduction of $10\log_{10}(768/2) \approx 25.84$ dB. Some tolerance in the result must be considered due to the statistical nature of the noise.

Comparison of the noise distributions within the matrices $\widetilde{\mathbf{T}}$ and $\widehat{\mathbf{T}}$ is also shown in fig. 4.37. The little structured texture of the noise after the refinement process is there due to the indirect decomposition made throughout the algorithm.

Repeating the simulation with much lower values for $|C^t|$ and $|C^r|$ demonstrates the unwanted biasing effect due to the low SNR available. For $|C^t| = |C^r| = 1$, i.e., SNR of 0 dB, the simulation result is shown in fig. 4.36(b), where the rest noise is no more Gaussian shaped. Although the result is still advantageous, it shows the necessity to start the refinement process with a significant SNR level.

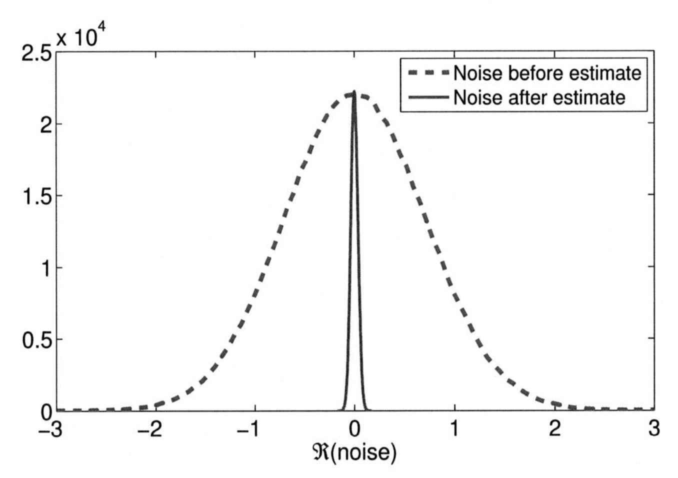


(a) Result using SNR of approximately 20 dB. Successful reduction in the noise power is achieved by a factor of 25.9 dB. Detailed comparison is shown in fig. 4.37.

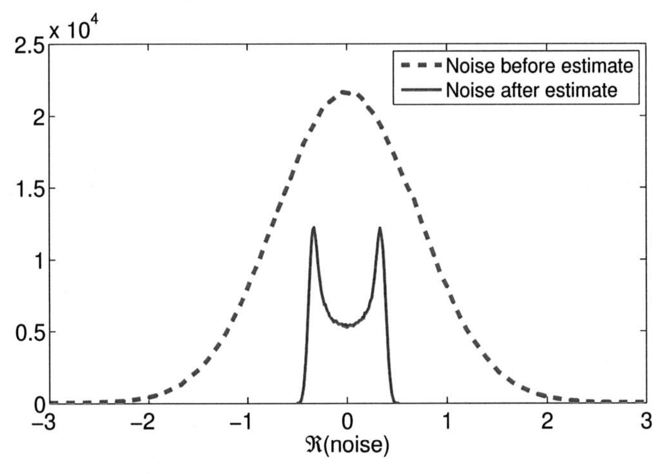


(b) Result using low SNR of 0 dB. Biasing effects are clear in the resultant estimate due to the relatively high noise power.

Figure 4.36: Simulation results of calibration matrix refinement using different SNR values. The real part of the noise before and after the refinement process is shown.

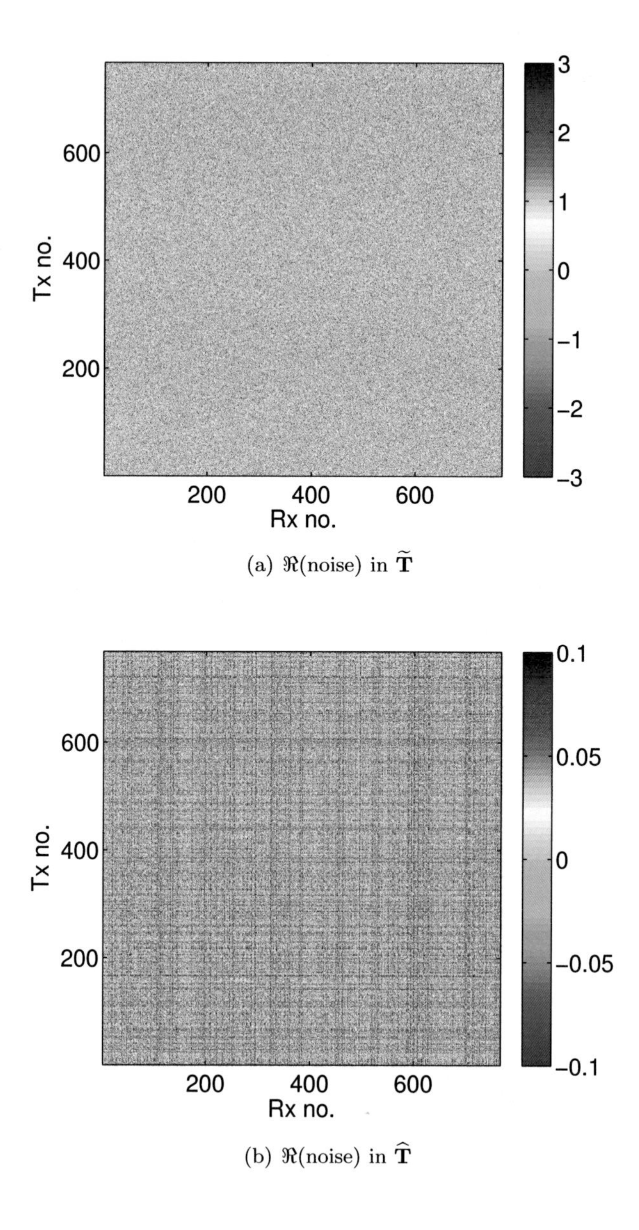


(a) $\Re$(noise) in $\widetilde{\mathbf{T}}$

(b) $\Re$(noise) in $\widehat{\mathbf{T}}$

Figure 4.37: The real part of the noise content before and after the calibration matrix refinement algorithm using simulation with SNR of approximately 20 dB. The different scaling demonstrates the successful refinement. The associated PDFs are also shown in fig. 4.36(a).

## 4.7 Conclusion

Calibration of multistatic imaging arrays has been discussed and analyzed thoroughly. A suitable calibration procedure based on two error model has been introduced, along with discussions on many practical aspects influencing the quality of the calibration. Calibration of imaging arrays can be disturbed by statistical errors as well as systematic ones. The statistical ones result mainly from thermal noise in the measurement chains, whereas the systematic ones arise mainly from mechanical misalignment causing mismatch between the considered expected geometry and the actual one.

The calibration procedure requires a good reflector with well-known reflection coefficients. Thus, for a large array operating in close range, a problem of proper characterization of the reference reflector has been recognized. Using electrically large spheres is suggested, however no known solution for its bistatic reflection coefficient at close range was available. Therefrom, a dedicated geometrical optics solver has been introduced to perform efficiently in the proposed problem. Accordingly, an analytical description of the specular point is essential for the performance of the solver. A suitable approximation has been introduced and proved to satisfy the accuracy required. In addition, care has been taken in calculating polarization mismatching due to the reflection on the curved surface of the sphere. The solver has been evaluated at the millimeter-wave range against the physical optics method and various experimental verifications, in all proving adequate quality and good efficiency.

Systematic errors in the calibration have been investigated, and their influences on image production have been examined. Deformation in reference phase plane can cause considerable image distortion and produce defocusing effects. Nevertheless, it has become evident that mechanical tolerances comparable to the used wavelength while positioning the calibration object are allowed. This can ease the calibration procedure a lot and thus avoids the costs of sophisticated mechanical alignment. A technique for geometry error correction has been discussed, thus suggesting an algorithm to enhance calibration quality.

Statistical errors caused by thermal noise have been addressed with a method for the refinement of the calculated transfer functions. It is the particularity of large multistatic arrays to offer high level of redundant information via the calibration procedure. This information is thus correlated efficiently to deduce a better estimate of the transfer functions. Enhancement of several tens of dBs in their SNR is proved achievable.

All these efforts are considered for an accurate calibration of multistatic imaging arrays. They should be hence combined in a way appropriate for the application in question.

Chapter 5

# Demonstrator

The imaging capability of planar multistatic arrays and their applicability for fully electronic imaging are demonstrated in this chapter. A demonstrator is conceived based on the array architectures discussed in sec. 2.8.4 to operate with digital-beamforming (DBF) technique in the millimeter-wave range [91, 112]. Its hardware components have been designed and fabricated in collaboration with Rohde & Schwarz, Infineon Technologies, and University of Erlangen-Nuremberg, based on the results presented in the previous chapters. Access to the E-band is achieved through a dedicated chip-set fabricated with silicon-germanium (SiGe) technology [97]. The demonstrator includes a total of 1472 antennas, which are populating an aperture area of approximately 50 cm times 50 cm. Coherent measurement acquisition in this frequency range and over such a huge number of channels is very challenging, therefore a lot of attention was taken during the system design to ensure high signal stability and quality. Therefore, heterodyne reception and continuous-wave stepped-frequency signals are utilized in order to achieve the intended measurement quality. Fig. 5.1 shows photographs of the demonstrator front-end part.

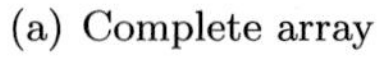

(a) Complete array

(b) Single cluster

Figure 5.1: Fully electronic multistatic imaging system with 1472 antennas placed in 16 clusters. An euro coin is used for size comparison.

In the following sections, the array design is discussed, followed by a description of the hardware architecture and the measurement acquisition. Afterwards, the signal processing and the system calibration are explained, followed by a presentation of the resultant microwave images.

## 5.1 Array design

The same array topology as in fig. 2.13(b) is adapted for the practical realization of the demonstrator. Identical $4 \times 4$ clusters are used, each with 46 Tx and 46 Rx antennas, i.e., 23 antennas per antenna line. The antenna lines are alternately displaced to a zigzag shape in order to ease the integration of the electronics and the antenna structure underneath. The antenna spacing $\Delta a$ is equal to 3 mm, which gets to increase in the effective aperture to $3\sqrt{2}$ due to the zigzag shape. Fig. 5.2 illustrates the geometrical details of a single cluster, and the shape of the effective aperture caused by the zigzag shape of the antenna lines. The spacing between the clusters follows the same geometrical relations as in fig. 2.20; accordingly, a cluster raster of 138 mm is used.

For proper imaging performance, it is essential that the antenna delivers wide opening angle hence low antenna gain, a stable phase center over angle and frequency, low side-lobe level, high suppression of the cross-polarized component, and high signal bandwidth. The used antennas in the demonstrator are all co-polarized with 45° polarization relative to the array coordinates. This allows using the same antenna structure on both, the Tx and the Rx lines by simple 90° rotation. The antenna structure consists of four layers utilizing a linearly polarized balanced-fed patch-excited horn antenna type, which was introduced in [113]. The antenna feed is a differential line connected to a dipole located in the same layer, all placed over a ground layer. The dipole hence radiates through two slots in the next layer, which then excite a patch in the upper most layer. Moreover, The patch radiates through a horn antenna, which helps in achieving a stable antenna phase center and in shaping the antenna pattern. The antenna structure and its co- and cross-polarized radiation patterns are illustrated in fig. 5.3. The antenna exhibits a gain of approximately 8 dBi in the co-polarized component and a good suppression of the cross-polarization one.

The array is intended to operate around 75 GHz, as the exact frequency sweep can be chosen by the signal source. For multistatic arrays with isotropic antennas, alias-free imaging is unconditionally achieved with $0.5\lambda$ spacing [24]. Here, the antenna spacing corresponds to $0.75\lambda$ at 75 GHz instead. Following the available opening angle of the antennas, the spacing is relaxed in order to reduce the system complexity and cost.

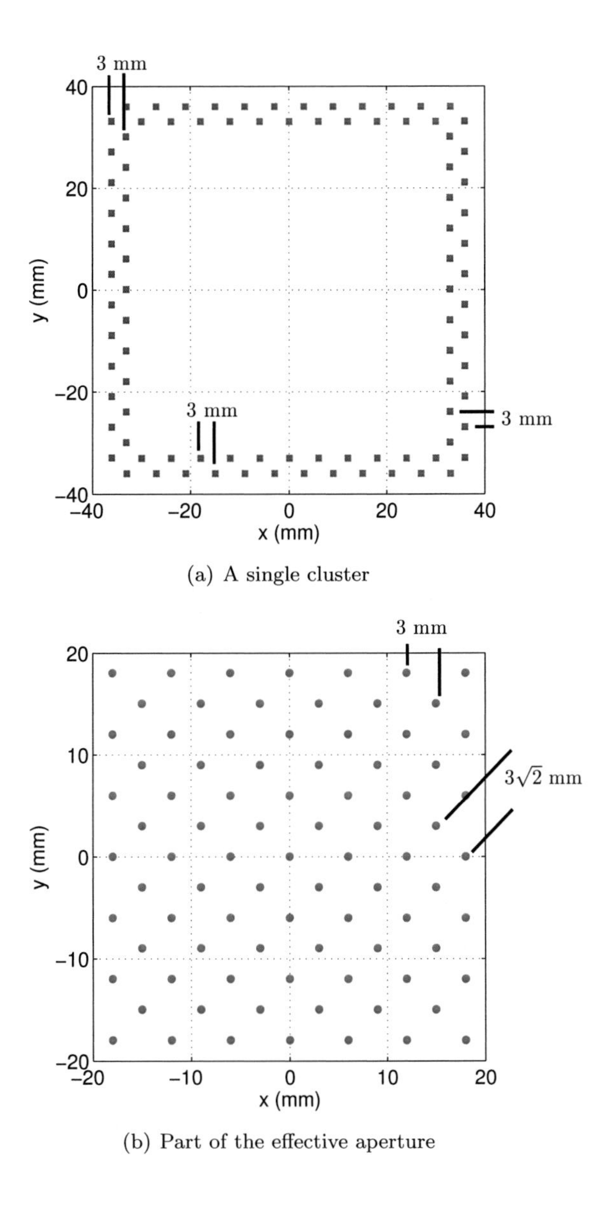


(a) A single cluster

(b) Part of the effective aperture

Figure 5.2: Geometry of a single cluster used in the demonstrator. A total of 46 Tx and 46 Rx antennas are integrated. The Tx antennas are placed horizontally, and the Rx ones vertically with a spacing of 3 mm. The antenna lines are displaced to a zigzag shape with 3 mm. The resultant arrangement in the effective aperture is also shown.

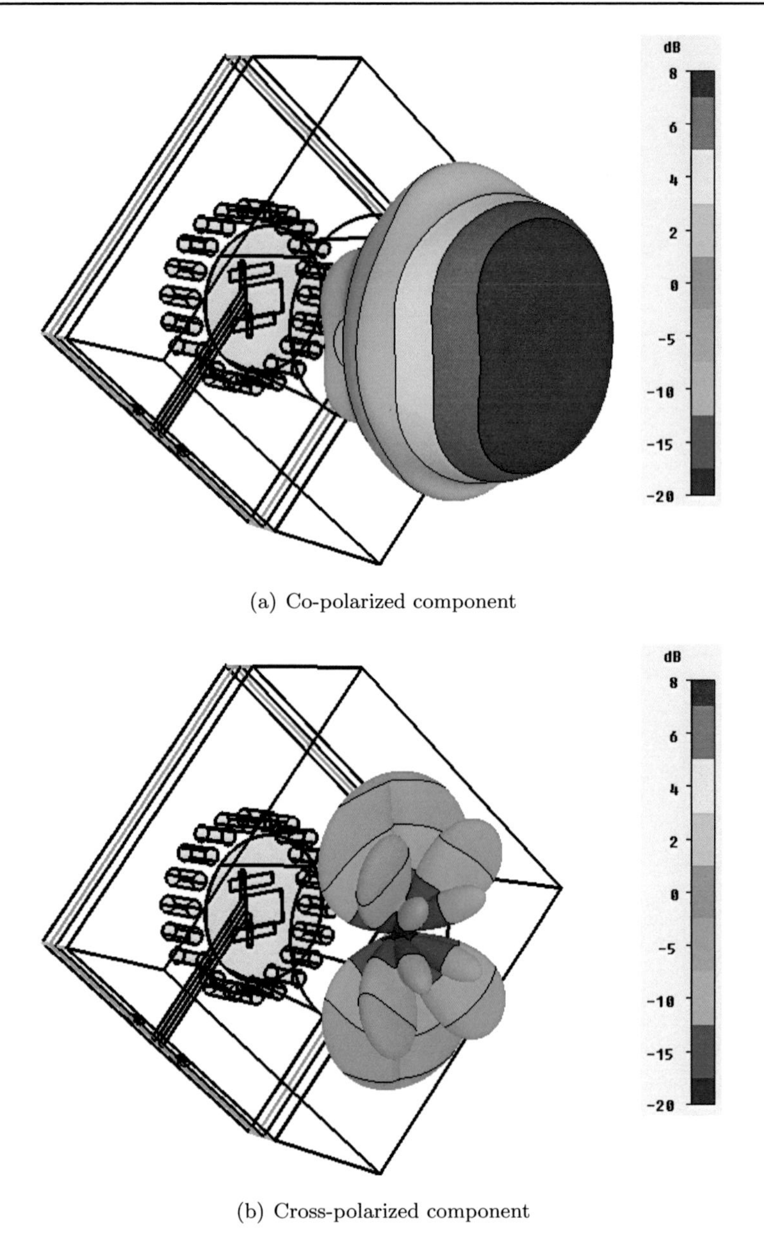


(a) Co-polarized component

(b) Cross-polarized component

Figure 5.3: Illustration of the antenna structure and its radiation pattern at 75 GHz showing the radiation quality for the co-polarized component and the good suppression of the cross-polarized one [113]. Patterns are simulated using a commercial 3D electromagnetic field simulator [114].

## 5.2 Hardware description

The imaging system must collect the reflection data coherently over all Tx-Rx combinations and all frequencies of operation. Accurate phase measurements are essential for the synthetic focusing afterwards. The system comprises a total number of 768 Tx channels and 768 Rx channels, whereas each of the 768 channels are distributed to 736 antennas and 32 reference channels, i.e., two reference channels per cluster. In order to ensure phase coherence allover the system, signal generation is made centrally and then distributed to the front-ends. Setting the source to a certain frequency, each transmitter illuminates the imaged object once, and the reflected wavefront is sampled by the receivers. The transmitters must be switched sequentially for the data collection, however the receivers can operate in parallel. The reflection data is collected for each Tx-Rx pair and at each selected frequency, forming a three-dimensional data volume of $N_t \cdot N_r \cdot N_f$ complex value. Heterodyne receivers are used to allow for a high dynamic range. In this realization, 48 receiver channels can be sampled in parallel, and hence the receivers are multiplexed 16 times in order to collect the whole data volume [112]. The collected data goes then through the system error correction procedure. Once corrected, the data is weighted as discussed in sec. 3.3. Lastly, image reconstruction is made using a numerically optimized space domain reconstruction based on the formulation in sec. 2.6.1.

Fig. 5.4 illustrates the block diagram of the imaging system. Utilizing heterodyne reception, two coherent signals per frequency step must be generated, namely the radio frequency (RF) and the local oscillator (LO) signals. A synthesizer generates both signals at one fourth of the intended frequency. These signals are then distributed and quadrupled inside each cluster. The RF signal is distributed to the Tx channels, and the LO signal to the Rx channels. The mixing products in the intermediate frequency (IF) is at 15 MHz. These signals are digitized and processed for magnitude and phase information using analog-to-digital converters (ADCs), which run synchronously with the signal source. Signal processing, hardware control, IF signal digitization, system error correction, image reconstruction, storage, and display are all performed by an industrial computer attached to the array hardware. This computer is customized based on a commercial PXI Express (PXIe) system, which includes PXIe controller, six 8-channel ADC units, and a programmable control card which contains the real-time capable parts of the control state machine. Fig. 5.5 shows photographs of the demonstrator while being integrated on a a movable platform to ease transportation.

In the following, detailed discussion about the system realization is presented.

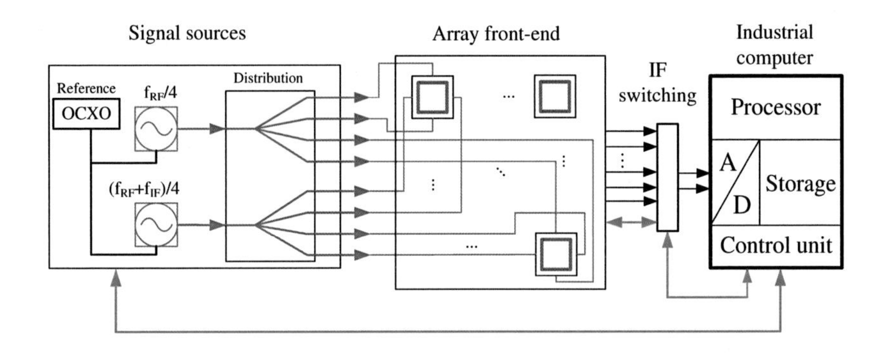


Figure 5.4: Hardware block diagram of the demonstrator illustrating the main signals inside the system.

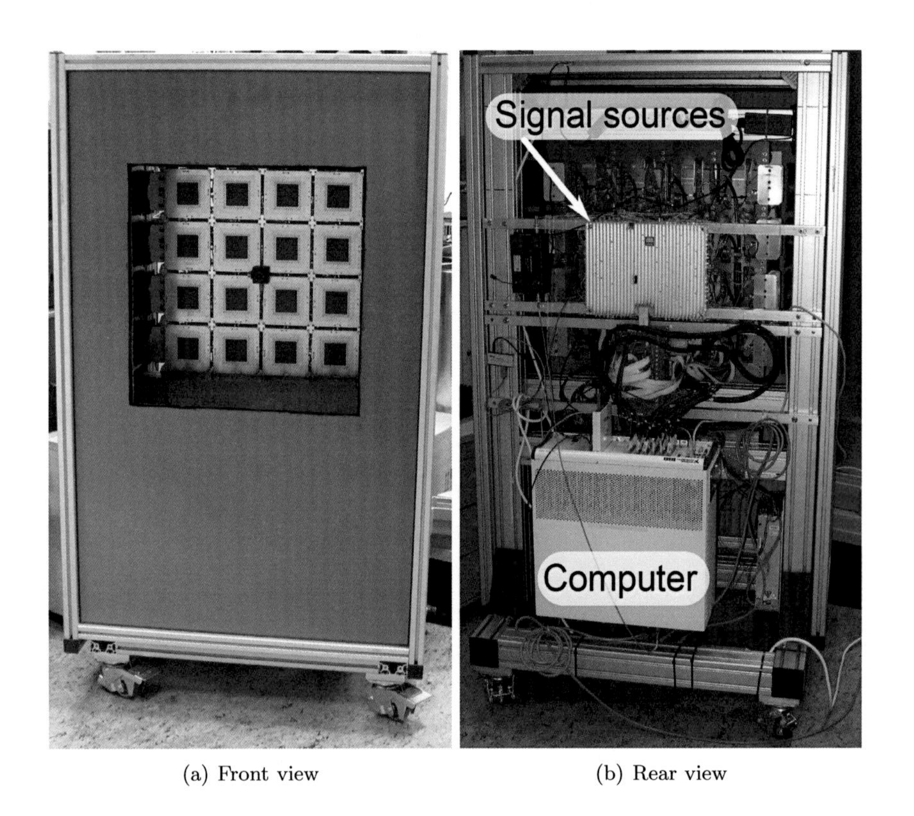


(a) Front view (b) Rear view

Figure 5.5: Photographs of the demonstrator while being integrated on a movable platform to ease transportation.

### 5.2.1 Signal sources

A commercial unit for signal sources has been developed and optimized by Rohde & Schwarz for this imaging purpose. The sources are prepared to generate the RF and LO signals using two direct digital synthesizers (DDSes) operating from 130 to 170 MHz, which are followed by seven multiplier stages. They are hence able to cover a frequency sweep range from 17 to 21.5 GHz, i.e., leading to a corresponding imaging frequency range from 68 to 86 GHz, which covers the frequency band of operation and beyond. This is a compromise to ease signal distribution to the front-ends, where the signals are quadrupled to the 80 GHz range inside the chips. Thus, the total multiplication factor to the target frequency is 512, which hence leads to strict specifications on the frequency reference source in order to keep sufficient signal quality at the target frequencies. Therefore, the clock of the two DDSes is derived from a highly stable oven-controlled crystal oscillator (OCXO). DDSes feature fast switching, thus being beneficial to reduce the total measurement time. Targeting an IF frequency of 15 MHz for the imaging system, a frequency difference between the two signal sources is selected to be equal to one-fourth the IF frequency, namely 3.75 MHz. The LO signal is chosen to be higher than the RF one, as illustrated in fig. 5.4.

Additionally, the distributed signals must be generated with sufficient power in order to derive the chips in the clusters. The two signals are distributed using Wilkinson dividers and are amplified to 32 output ports for the RF signal, and another 32 ones for the LO signal. The two distribution networks are integrated with the DDSes in one unit. This ensures good thermal coupling between all amplifiers used along the distribution networks, hence enhance phase stability of the system. An output power of approximately 20 dBm per port is availble.

### 5.2.2 Array front-end

The Array front-end consist of 16 clusters. Each cluster contains 46 Tx antennas, 46 Rx antennas, and two reference channels. The cluster integrates 12 Tx chips, 12 Rx chips, signals distribution, and the 92 antennas. Fig. 5.6 shows the RF and LO connections to a single cluster and the integration of its inside components [115]. A tailored printed circuit board (PCB) manufacturing process was established by Rohde & Schwarz to miniaturize the cluster components. Each side of the cluster contains either six Tx or Rx chips, with an RF or LO input port respectively. Signal distribution for the six corresponding chips is made within the cluster itself, which is realized by a 1-to-3 Wilkinson divider followed by three 1-to-2 Wilkinson dividers. Each chip integrates four channels with dif-

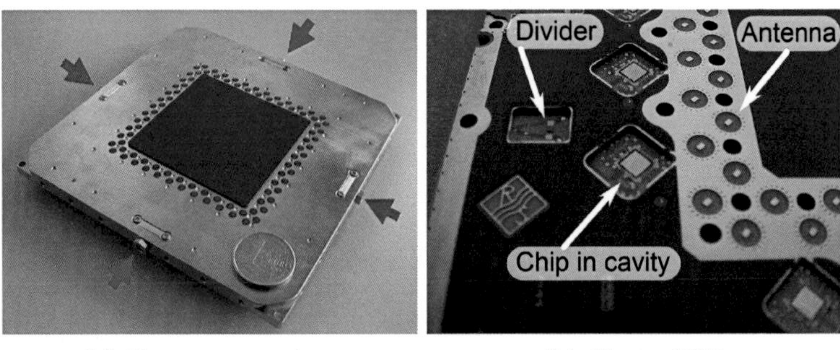


(a) Cluster connections (b) Cluster PCB

Figure 5.6: Cluster connections and inside components. Red arrows point the RF input for the Tx chips, and the blue ones for the LO input to the Rx chips. Control and IF signals are connected on its backside. A total number of four 1-to-3 Wilkinson dividers, twelve 1-to-2 Wilkinson dividers, twelve Tx chips, twelve Rx chips, two reference channels, and 92 antennas are all integrated in one multilayer printed circuit board.

ferential lines to the antennas. Fig. 5.7 illustrates their functional blocks (reproduced from [97]). The chips are manufactured by Infineon Technologies using an industrial automotive SiGe:C bipolar process, and are optimized to offer large bandwidth with less output power as compared to typical automotive chips [116]. The deployed chip-set operates from 68 to 85 GHz with an approximate output power of 0 dBm and a receiver conversion gain of approximately 23 dB [97]. Each chip allocates an area of 4 mm$^2$.

The cluster surface is formed by a gold plated aluminum cover, which integrates the horn part of the antenna and the connections for the RF and LO signals. The control and the IF signals are all routed through the backside of the cluster, which connects to the control board with a switch matrix. The inner part of the outer surface of the cluster is prepared with an absorber sheet in order to reduce the cross-coupling between antennas and the standing waves between the imaged object and the array surface.

### 5.2.3 Control units

Each cluster is connected to a control board, which offers the interface to the main control unit in the industrial computer as well as a set of multiplexers with a switching matrix [117]. The 48 IF signals of each cluster are amplified and multiplexed to three IF outputs, i.e., 16-to-1 multiplexing. The main control unit thus controls the 16 control boards

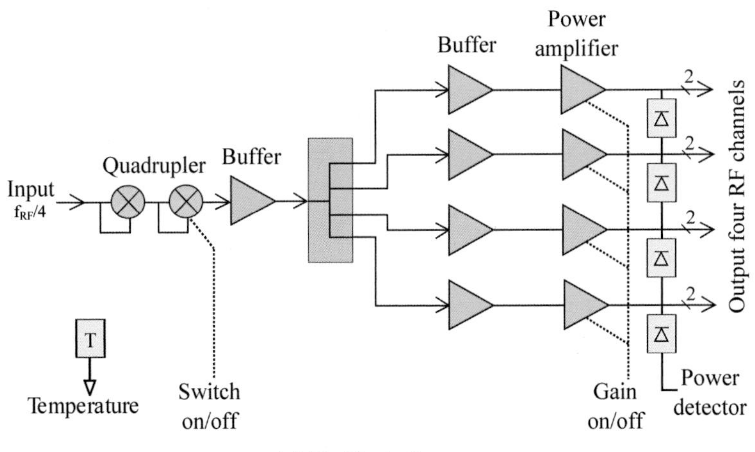


(a) Tx block diagram

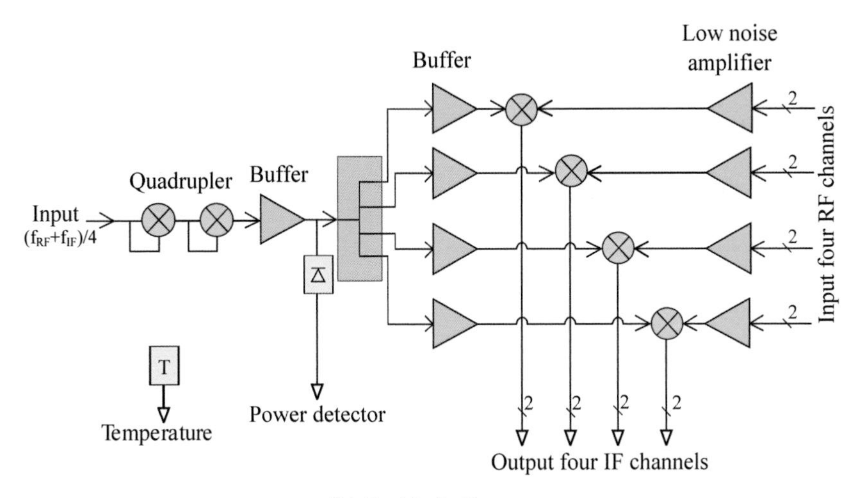


(b) Rx block diagram

Figure 5.7: Dedicated four channel analog front-end chips used in the demonstrator [97]. The connections to the antennas are made with differential lines. The Rx channels operate in parallel and deliver the IF signals in differential form to enhance channel isolation. The Tx chip operates sequentially, thus only one Tx is switched on per time. This is controlled using the gain on/off control signal. When no Tx channel is active, the quadrupler is switched off to ensure complete absence of any leakage signals. Power detection at relevant positions as well as the chip temperature can be monitored using integrated sensors.

as well as the signal sources. The 48 ADCs integrated in the computer platform are also controlled and kept running synchronous with the reference source during measurement acquisition. Each ADC unit is running with a sampling rate of 60 MSps, hence can sample exactly four samples per a period of the IF signal.

### 5.2.4 Measurement acquisition

To record the data for image generation, one Tx illuminates the scenario after another at each frequency, while the reflection data is recorded by the 48 ADCs after being multiplexed. A flowchart of the applied steps is summarized in fig. 5.8. The IF signals are continuously digitized and stored in the integrated memory of the ADC unit during the complete measurement acquisition, even during switching processes when the IF signal is invalid. After the measurement acquisition is finished, the stored IF samples are processed as follows: the valid samples are selected from the memory, then digitally downconverted and filtered using a 16-tap low-pass digital filter with a power bandwidth of 1.875 MHz. Applying this procedure to all Tx-Rx-frequency combinations results in a 3D data structure of complex-valued wave quantities, which is stored for further signal processing steps.

The measurement acquisition time using $N_f$ frequencies is given approximately by

$$N_t \cdot \frac{N_r}{N_{\mathrm{ADC}}} \cdot N_f \cdot \frac{N_s}{f_s}\,, \tag{5.1}$$

where $N_{\mathrm{ADC}}$, $N_s$, and $f_s$ denote the available number of ADCs, the number of samples measured per channel and frequency, and the sampling rate achieved by the digitization hardware [112]. $N_s$ is selected to be equal to 24, in which 16 samples are used for determining the wave quantity and another 8 samples are left due to settling time of the IF chain including multiplexers and amplifiers. The settling time of the signal sources is neglected here due to the very fast DDS switching time relative to the IF chain. For $N_t = 768$, $N_r = 768$, $N_{\mathrm{ADC}} = 48$, $N_f = 64$, $N_s = 24$, and $f_s = 60$ MSps the measurement acquisition time is 315 milliseconds. The selection of $N_f$ depends on the required alias-free imaging range. Assuming equidistant frequencies, the classical radar far-field relation approximately applies, in which $N_f$ is related to an alias-free imaging range $d$ by

$$N_f = \frac{2d}{c} \cdot \left(f^{\max} - f^{\min}\right) + 1\,. \tag{5.2}$$

It is noticeable to mention that during the measurement at each single frequency, coherence between signal sources and the ADC clock must be achieved, and the recorded samples must be aligned precisely in time to the corresponding transmitter.

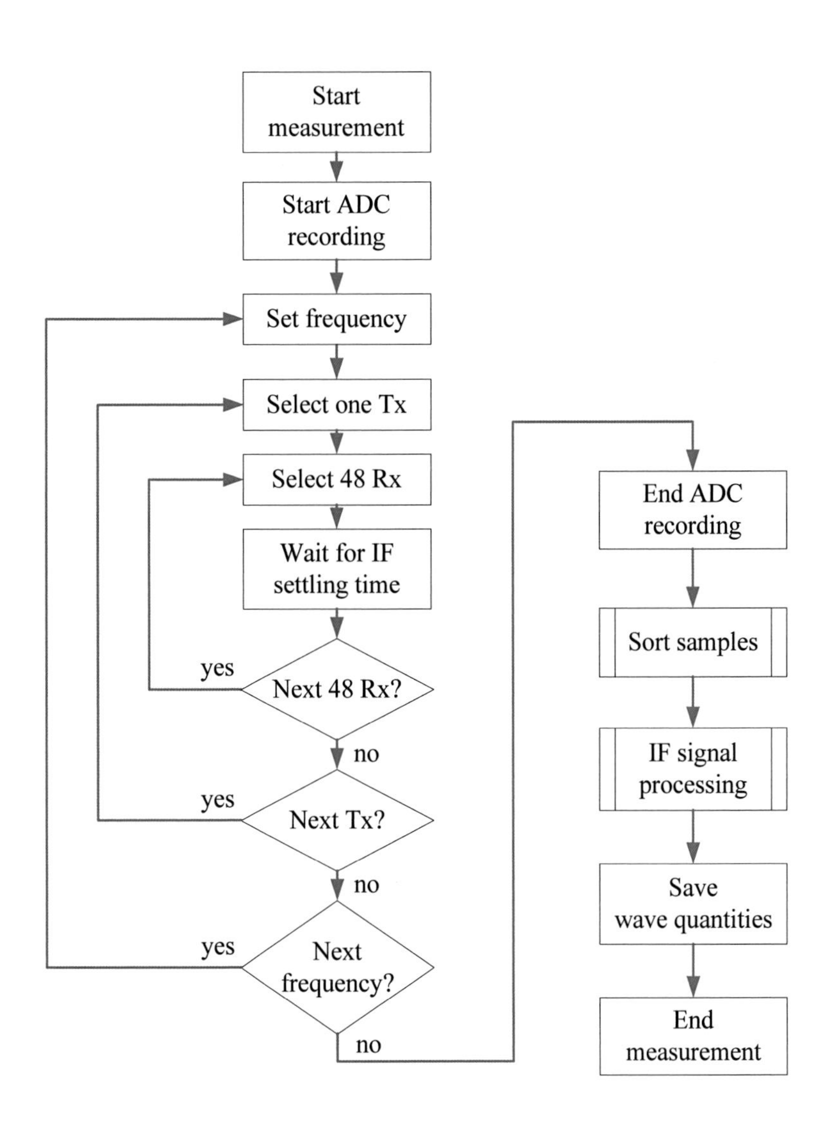


Figure 5.8: Flow chart of the measurement acquisition used in the demonstrator.

## 5.3 Signal processing

The processing of the three-dimensional reflection data undergoes various steps to be prepared for image reconstruction. Fig. 5.9 illustrates the sequence of these steps. At the first step, the reflection data has to be referenced using the integrated reference channels in each cluster in order to get phase stable and reproducible signals. In each cluster, one of the two available reference channels is used as a reference signal for all transmitters within the same cluster. The reference channel includes a 40 dB attenuator pad in order to ensure linear operation of the receiver. As the reference channel is treated similar to an antenna, it is not measured continuously, however only when its transmitter channel is active. Thanks to the phase stability of the signal sources, the produced time delay between the reflection measurement and the reference channel measurement causes reproducible phase difference, which is consequently corrected during the system error correction. Once the data is referenced, the system error correction is applied as in eq. 4.2, which hence yields to the actual reflection data from the imaged object.

Afterwards, the data is filtered using time gating technique in order to eliminate any residual cross-coupling within the array front-end and to reduce the clutter signals from unwanted objects prior to image reconstruction. For this, the reflection data is converted to time domain using inverse discrete Fourier transform at each Tx-Rx combination. A gating function is then applied and the signals are transformed back to frequency domain using discrete Fourier transform. Fig. 5.10 illustrates the time domain histogram of an example measurement. The histogram represents the statistical distribution of the reflected signals along range direction. In this case, the main reflections exist around the 70 cm distance, and hence the time gate is selected from 50 to 90 cm. Some of the reflection signals are present outside this interval, which originate either from multiple reflections with the array surface itself, residual cross-coupling within the array front-end, or due to clutter from surrounding objects.

Having the reflection data now corrected, aperture weighting can be applied. Illumination equalization is performed followed by aperture smoothing according to eq. 3.4. This is a necessary step for the correction of the specular reflections produced on the smooth parts of the imaged object. For applications targeting only diffuse objects, the equalization step might be dropped in the processing. Here, a Kaiser window with shape factor of two is used to smooth the aperture, presented in fig. 3.8. Finally, the reconstruction algorithm takes place for focusing. Space domain focusing with numerically optimized implementation is applied. Processing time takes a few minutes depending on the reconstruction parameters and results in a fully focused 3D complex-valued image.

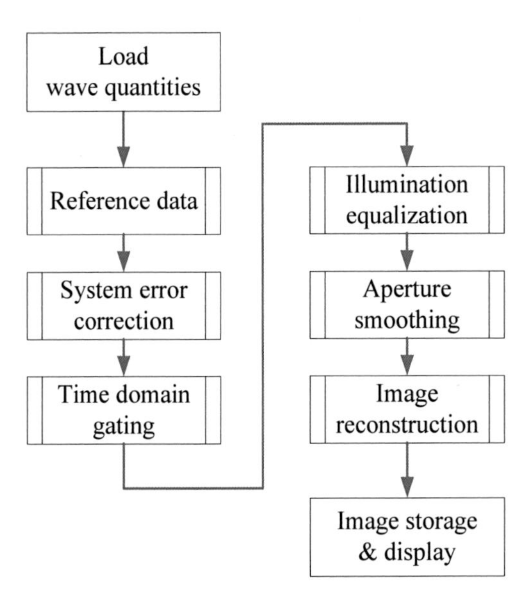


Figure 5.9: Flow chart of the signal processing steps used in the demonstrator to reconstruct the 3D images from the measured wave quantities.

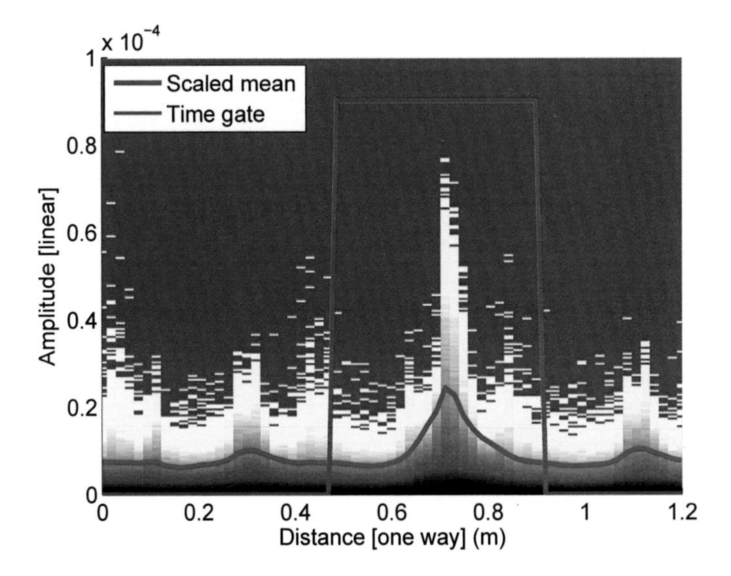


Figure 5.10: Time domain histogram of an example measurement illustrating signal distribution along range direction. Time gating is applied to suppress unwanted signals.

## 5.4 System calibration

The collected reflection data must be corrected in order to set the phase reference of the imaging system directly on the plane of the phase centers of the Tx and Rx antennas for all frequencies of operation. Thus, the purpose of the calibration procedure is the determination of the matrices $T$ and $\Gamma_{cc}$ based on the method discussed in sec. 4.2. The two required calibration measurements, i.e., match and offset short, are realized using pyramidal absorbers and a metallic hemisphere. As discussed before, the match measurement helps to suppress the cross-coupling within the array, whereas the offset short measurement is used to correct the phases of the Tx and Rx channels and to equalize their response over frequency. The multiple reflections between the array surface and the imaged object is partially reduced by the absorber sheets attached to the array surface and through the time gating step. As no time gating can be achieved during the calibration procedure itself, i.e., channels phases are not yet corrected, care is taken in the selection of the offset short standard. Using a hemisphere, the majority of the signals reflected form the hemisphere will reflect back from the flat array surface in a direction away from the hemisphere. Additionally, the signals exposed to a second reflection on the hemisphere will be much weaker in magnitude due to signal diffusion on its curved surface, sec. 4.4.3.2. Fig. 5.11 shows the metallic hemisphere used to apply the calibration procedure. The calibration measurements are referenced to the reference channels the same way discussed above; and are additionally enhanced in quality using 100 times averaging. In order to apply eq. 4.5 and eq. 4.6, a simulation of the sphere reflections $\Gamma_{sim}$ must be known. For this, a simulation based on the method introduced in sec. 4.4.3 is performed for a sphere radius of 15 cm and a distance of 65 cm to the array. The knowledge of antennas radiation pattern, polarization, and phase center presented in sec. 5.1 are also considered.

For closely spaced Tx-Rx pairs, the cross-coupling could lead to compression in the Rx channel. Once happened, the channel does not behave linearly, which causes the calibration procedure to fail. Therefore, these pairs are masked out intentionally to avoid inconsistencies in the image reconstruction later on. Due to the array geometry, only less than 0.1% of the collected data volume is considered invalid. Fig. 5.12 illustrates the antenna's indexing inside the cluster along with an example for measured data after system error correction for four neighboring clusters. The object reflections are seen distributed over the aperture. The masked channels in the diagonal direction correspond to corners of the clusters. The horizontal and vertical dark lines correspond to the reference channels and to two damaged Tx channels in cluster one, which are all masked out from the following signal processing steps.

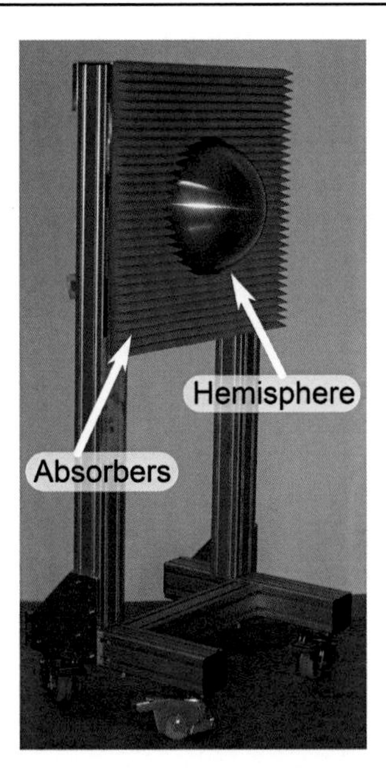


Figure 5.11: Calibration object used as an offset short, namely a hemisphere of 15 cm radius surrounded with absorbers. The center of the hemisphere is positioned on a distance of 65 cm centrally to the array.

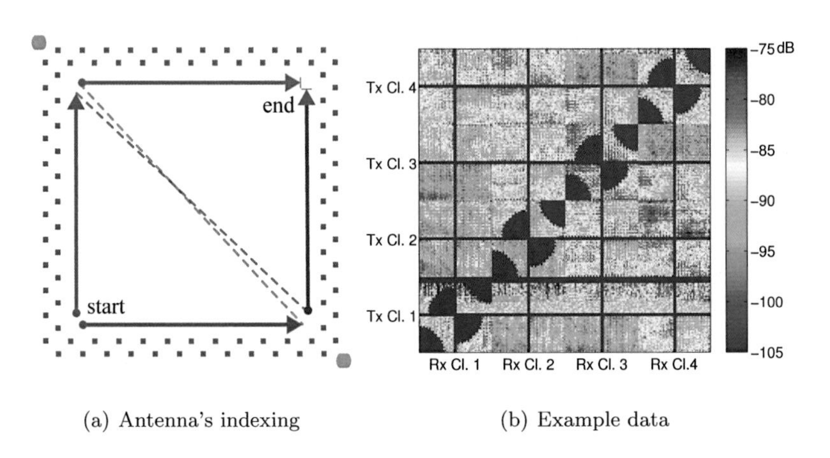


(a) Antenna's indexing

(b) Example data

Figure 5.12: Illustration of the Antenna's indexing used in cluster. The green patches show the position of the two reference channels. An example data after system error correction for four neighboring clusters is also shown. The color codes the maximum value over frequency. The masked positions in dark blue include cluster corners and reference channels.

## 5.5 Imaging results

Imaging with the demonstrator is proved using different scenarios. First, the lateral resolution achieved by the system is verified in order to examine the successful coherence along the aperture. Any lack of spatial coherence would cause loss in the lateral resolution. For this, a U.S. Air Force (USAF) chart is used for testing. The chart includes precise opening slots made in a metal sheet of 1 mm thickness. The slots are made in two orientations, i.e., vertical and horizontal. The test chart is positioned centrally to the array and parallel to the aperture on a distance of 520 mm. The imaging results are shown in fig. 5.13 after using a frequency range from 68.5 to 82 GHz in 91 steps. Magnitude and phase image at 520 mm distance are present, along with a a schematic drawing of the chart dimensions. A phase ramp along the chart surface is observed, which results from being slightly tilted relative to the parallel plane to the aperture. It demonstrates, however, the exactness of the image reconstruction represented by the quality of the phase information. For determining the theoretical lateral resolution expected from the system, the formula in eq. 2.74 is applied using the values $f^{\max} = 82$ GHz, distance $L = 520$ mm, and aperture width $D = 486$ mm. This yields a theoretical lateral resolution value at the center position of 2.16 mm. By examining the magnitude image in fig. 5.13, the separation between the opening slots in both directions is clearly achieved down to the 2.2 mm openings. Fig. 5.13(d) shows the magnitude values of trace crossing the 2.2 mm vertical slots marked in fig. 5.13(c). Separation greater than 3 dB is available, thus validates the formulations in sec. 2.7.2..

Next, the image quality is examined regarding artifacts. A test object is prepared including weak reflectors and a strong reflector, with the intention to examine the visibility of the weak reflectors in existence of the strong one. In case of strong side-lobe or lack of coherence, the visibility of the weak reflectors might get distorted. The used test object is shown in fig. 5.14(a), which consists of a bed of nails mounted with pyramidal absorbers in the background. The nails are metallic and are fixed to a regular grid of 10 mm distance, hence $10\sqrt{2}$ diagonally. Each nail is of 5 mm in diameter with an approximate radar cross-section of 50 dBsm. The USAF chart is added to the foreground as a strong reflector relative to the small nails. Fig. 5.14(b) shows a 3D rendering of the magnitude of the resultant image in linear scaling, while being prepared for a slanted view in the horizontal plane. All nails within the imaged volume are imaged successfully without significant artifacts; and the visibility of the nails and the USAF chart does not show any distortion. By comparing the brightness level of the USAF and the average brightness of the background empty voxels, better than 30 dB difference is achieved.

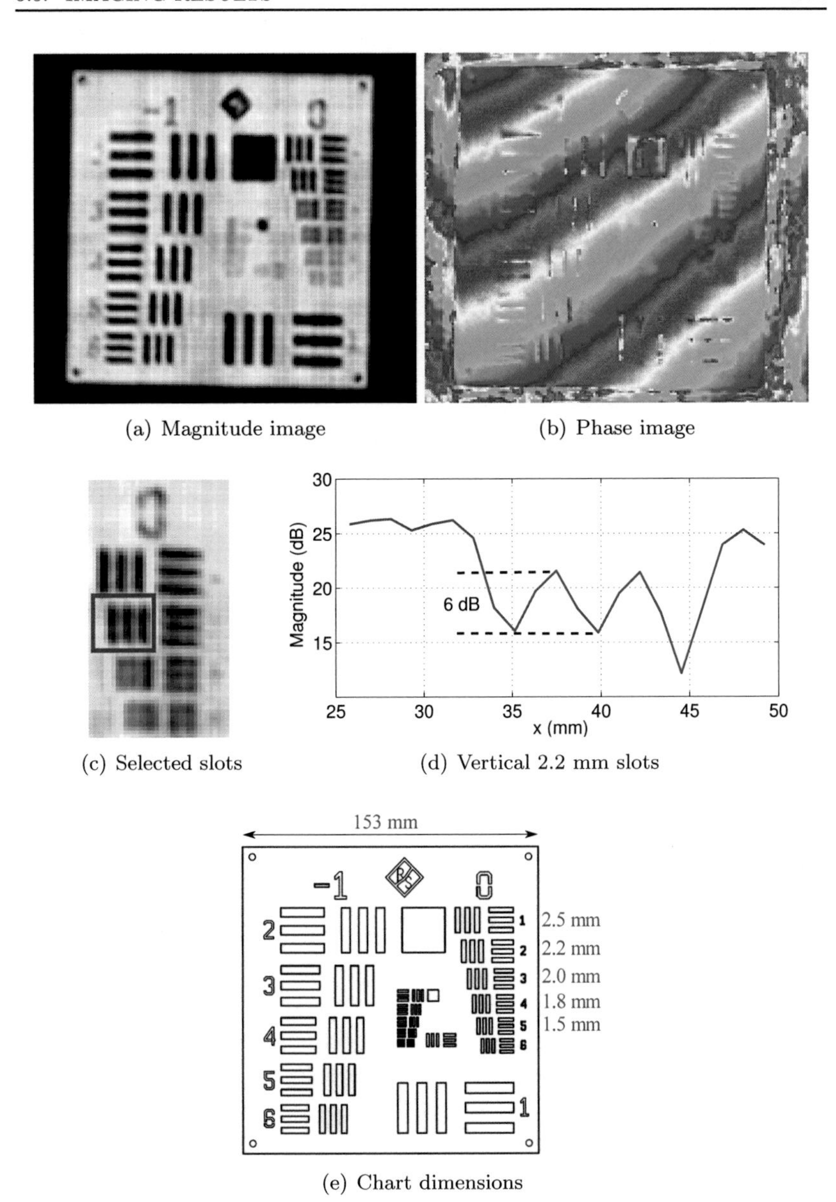


(a) Magnitude image

(b) Phase image

(c) Selected slots

(d) Vertical 2.2 mm slots

(e) Chart dimensions

Figure 5.13: USAF chart used to examine the lateral resolution achieved by the demonstrator. Being positioned on 520 mm distance parallel and central to the array, opening slots down to 2.2 mm are clearly resolved. Phase image shows the slight tilting of the chart and demonstrates the exactness of the phase information in the reconstructed image.

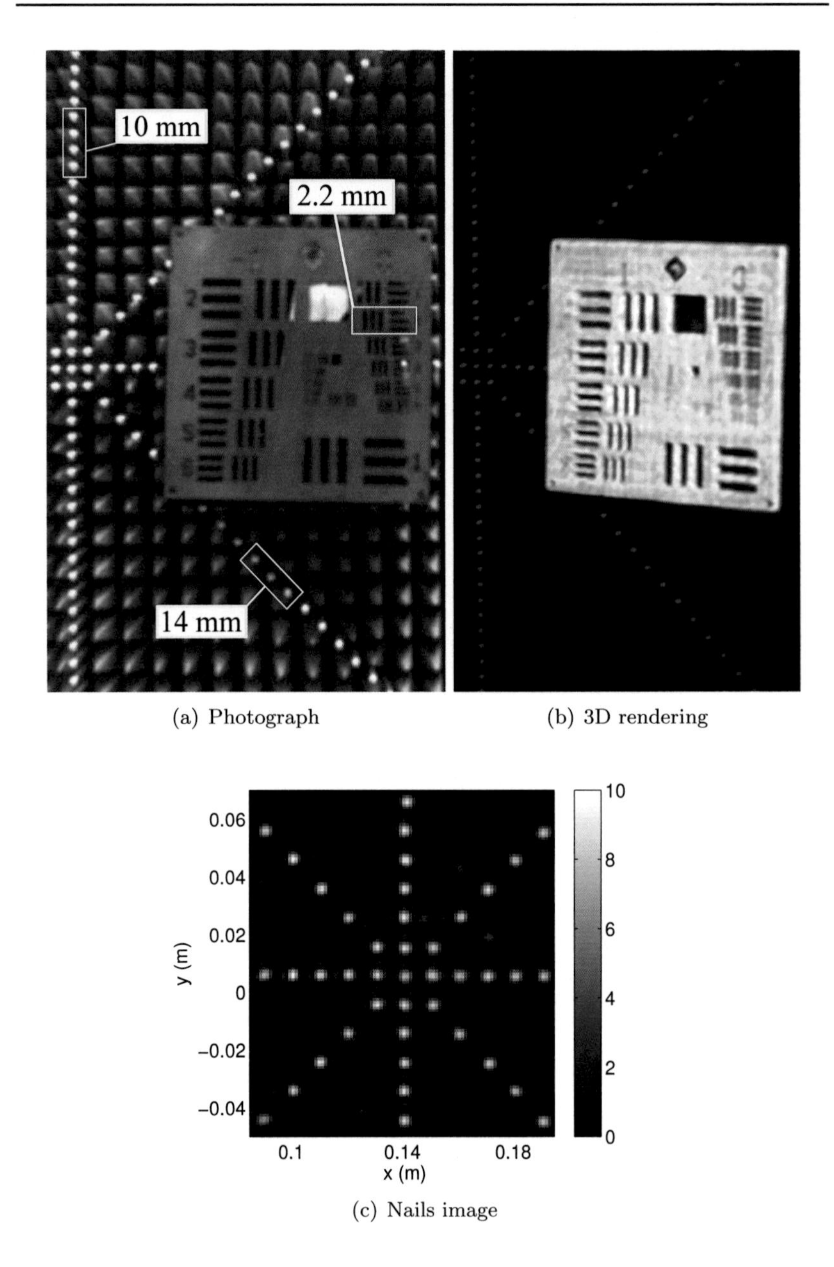


(a) Photograph (b) 3D rendering

(c) Nails image

Figure 5.14: Image of the USAF chart and a bed of nails test object prepared with pyramidal absorbers in its background. A 3D rendered view of the magnitude image is shown. Below, the image at the plane including the tips of the nails is present in linear scale.

Imaging humans is especially challenging. Fast measurement time is crucial to avoid blurring effects. In order to apply the synthetic focusing successfully, movements comparable to the wavelength during measurement acquisition must be avoided. Breathing, for instance, causes chest movement of roughly 10 mm/s. Thus, with 75 GHz illumination, i.e., 4 mm wavelength, the measurement time corresponding to a movement equal to one fourth the wavelength is 100 milliseconds. While imaging in reflection mode, change in range with one fourth the wavelength results in complete destructive addition with $\pi$ phase difference. Reducing the measurement time can be achieved here only by reducing the number of frequencies used during the measurement. By reviewing the relations addressed in sec. 5.2.4, the frequency range is chosen to be from 72 to 80 GHz with 65 steps. Additionally, the screened persons are requested to stand or sit still and to hold breath during the measurement acquisition in order to avoid blurring.

Fig. 5.15 shows the imaging result of a person carrying the USAF chart. The person sat approximately 50 cm from the array front-end. The associated time domain histogram of this measurement is the one shown in Fig. 5.10. The magnitude of the 3D image data is projected on 2D using the maximum value along range direction. The 2D image is prepared in dB scale and displaying with 30 dB dynamic range. Logarithmic scale is preferable in order to emphasize the weak reflections as well as the strong ones. The image is then colored according to the range information for each voxel, where the colors range from blue to red representing close to far distances relative to the array, respectively. Depth is calculated as the range position, where the maximum reflectivity takes place. Thus, the produced colored 2D image assembles the 3D information of the imaged object, where the intensity represents the magnitude of the maximum reflectivity and the color represents the depth information. The test chart was not oriented parallel to the array, which helps to emphasize the effect of the specular reflections. The surface portions points away from the array illumination effective aperture looks dark in the image. In spite of this, the edges are well visible due to the strong diffuse reflections they produce. It is to be noted that the microwave image is viewed using parallel projection, whereas the optical one is in perspective projection; and that both are measured simultaneously.

An important feature for microwave imaging is its ability to penetrate optically opaque surfaces. This is specifically beneficial for the application of personnel screening, which finds place at airport and critical infrastructure buildings. Two scenarios are presented next to demonstrate signal penetration through cloths. Fig. 5.16 shows imaging result for a person concealing a clasp knife. The knife is hidden behind a pullover and a shirt, however is clearly visible in microwave image with great details.

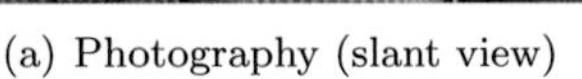

(a) Photography (slant view)

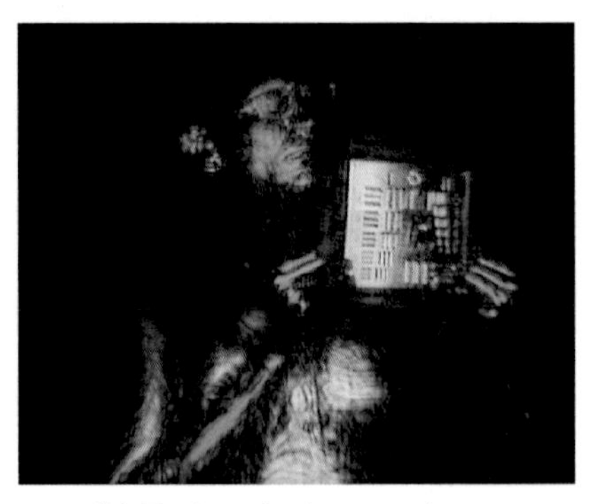

(b) Projected microwave image

Figure 5.15: Image of a person carrying USAF chart [118]. A frequency range of 72 to 80 GHz is used. Colors code depth information.

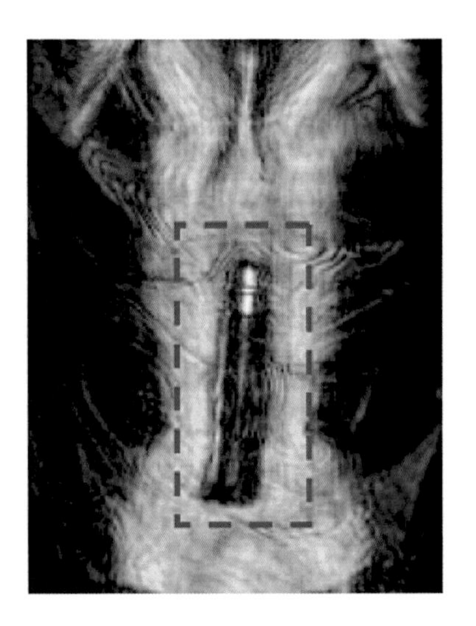

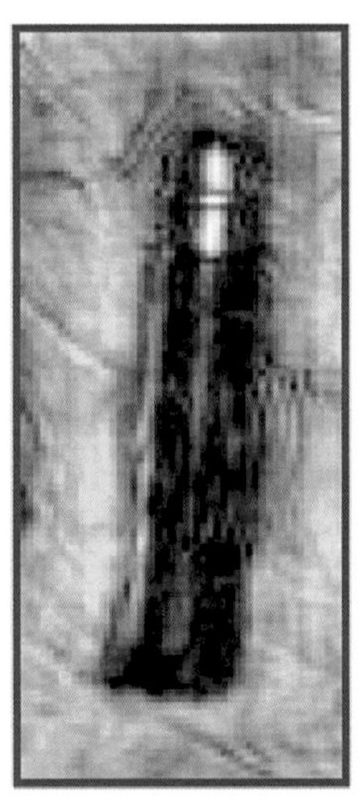

Figure 5.16: Image of a person concealing a clasp knife on his back and behind a pullover and a shirt [118]. Image is prepared as the projection of maximum intensity of the magnitude image in dB scale. Illumination effects are again observed at the sides of the image, where the skin vanishes due to the specular reflections. The diffuse scattering made by the clothing folds cause them to appear even where the skin is not visible.

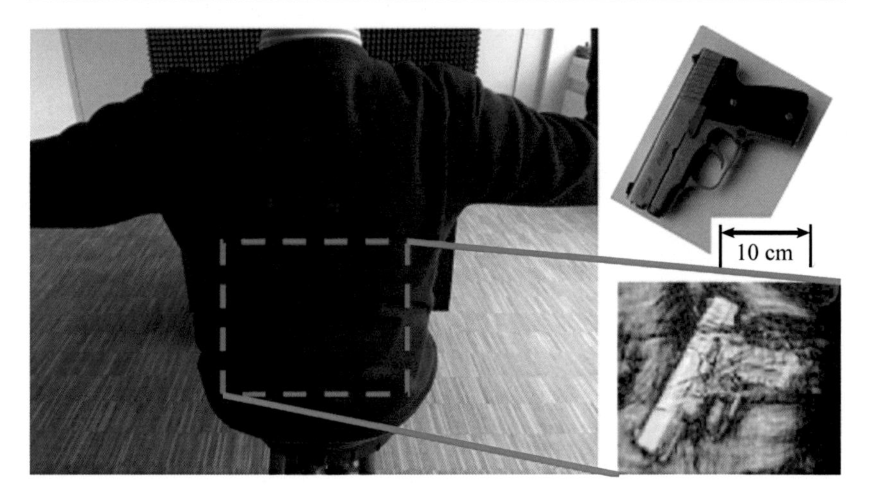


Figure 5.17: Image of a person concealing a small handgun behind a thick pullover [118].

In a similar preparation, fig. 5.17 shows an image of a concealed small handgun on the back of a person and behind a thick pullover. Optical images are also presented for illustration.

## 5.6 Conclusion

Fully electronic microwave imaging has been proved to be feasible with the proposed planar multistatic arrays. Digital-beamforming for synthetic focusing at close ranges has been demonstrated to be efficient in achieving imaging meeting design goals. Array design, resolution limits, calibration technique, and illumination effects have successfully met the theoretical discussions made in chapters 2, 3, and 4.

Measurement speed is recognized as a crucial factor in stabilizing image creation. Trade-off between the number of measured frequencies and the total measurement time must be considered depending on the application in question. Real-time imaging of humans is, however, possible with the demonstrator. Human skin is very reflective in the millimeter-wave range; and accordingly imaging of humans proves the validity of the illumination issues analyzed previously. Penetration capability of the millimeter-waves through cloths has been successfully demonstrated.

The demonstrator is highly configurable over large frequency range, and for various reconstruction parameters; and in addition is also made transportable and easy to calibrate and handle.

Chapter 6

# Summary

Microwave imaging with multistatic arrays features advanced properties. Planar multistatic arrays are specifically practical to realize and offer several enhancements compared to classical monostatic techniques. While operating at close ranges, image resolution and illumination quality enhance significantly in comparison with the operation in far-field. Range resolution, for instance, depends strongly on the imaging frequency. Besides the range resolution achieved by signal bandwidth, angular diversity offered by multistatic arrays considerably extends signal support in the spatial spectrum, thus leading to a consequent improvement of the range resolution. Unlike the conventional range radars, range focusing without signal bandwidth is possible with planar multistatic arrays. Therefore, in addition to the signal bandwidth, the imaging frequency becomes of particular importance for achieving high quality focusing. Moreover, phase information in the focused microwave images depends also on the imaging frequency. Phases can be used to extract features of the imaged object in the sub-wavelength level. Therefore, imaging at high frequencies, e.g., in the millimeter-wave and sub-millimeter-wave ranges, is favored to achieve three-dimensional microwave images of exceptional quality.

Dense multistatic arrays are hardly realizable, and, instead, sparse arrays are preferred in order to reduce hardware complexity, image acquisition time, power consumption, and system cost. Synthesizing a sparse multistatic array, which maintains the mentioned image quality and at the same time offer a realizable architecture, is very challenging. Based on the achieved understanding of the operation of planar multistatic arrays, several solutions have been presented to satisfy the imaging requirements. Hence, optimized sparse geometries proved to achieve these demands, while offering huge reduction factors in the total number of antennas. An array of one square meter aperture operating from 70 to 80 GHz has been synthesized. It comprises a total of 1536 Tx and 1536 Rx antennas, all grouped in 16 clusters. The reduction factor achieved is equal to 147 compared to a monostatic array of identical dimensions. The image quality of the array has been proven experimentally, showing an excellent agreement with the predicted performance. The introduced array topology is highly scalable, making it a valuable solution for many demanding applications.

When imaging objects of electrically smooth surfaces, illumination issues due to specular reflections arise. The association between the array geometry and the observed illumination quality in microwave imaging has been comprehensively discussed. Redundant array allocations in the effective aperture of planar multistatic arrays lead to inhomogeneous illumination. Therefore, an illumination equalization method is introduced to correct for this drawback. Additionally, the specular reflections have been proven to behave different than the diffuse ones through the focusing process, thus causing a phase mismatch limiting the maximum achievable brightness level. A quantitative analysis has been conducted to specify the theoretical limits for the brightness levels associated with multistatic arrays, i.e., corresponding to specular and diffuse reflection phenomena. Consequently, a suitable simulation technique has been introduced to predict efficiently the reachable illumination quality while imaging electrically large objects of complex three-dimensional surfaces. This simulation method has been experimentally verified and then utilized to determine the illumination on the human body for the typical application of personnel screening.

Imaging at microwave frequencies requires proper calibration of the imaging array. Synthetic focusing can only be achieved successfully when a determined reference phase surface exists. Bringing the introduced imaging concept to an actual realization has motivated the establishment of a realistic and yet efficient calibration procedure. In order to apply this calibration procedure, a reference reflector with accurately known complex reflection coefficients is demanded. For practical reasons, usage of electrically large spheres is recognized to be suitable to the underlying requirements. Interestingly, no analytical solution for the sphere near-field bistatic scattering is known. Therefore, a dedicated electromagnetic solver has been developed and successfully verified. Besides the system error correction applied to the imaging array, a method for the geometry error correction has been conceived and discussed. While imaging at the millimeter-wave range or beyond, precise positioning and alignment of large calibration objects become impractical. This imposes uncertainty in the calibration procedure and might considerably degrade the calibration quality. The introduced solution has been studied analytically and investigated numerically. In addition to systematic and geometrical errors, noise signals in the calibration measurements cannot be avoided. Thanks to the operation principle of multistatic arrays, a refinement process based on efficient matrix operations has been presented to optimally correlate the redundant information in the system error correction parameters and hence to reproduce the parameters with better signal-to-noise ratios.

A fully electronic demonstrator has been realized in order to verify the results achieved. It comprises a planar multistatic array with 736 Tx and 736 Rx antennas placed on an area of 50 cm times 50 cm. The array follows the previously introduced array topology, and hence verifies its scalability. The demonstrator operates from 68 to 82 GHz in a stepped-frequency continuous-wave mode. The design architecture is based on digital-beamforming technique, where image focusing was achieved synthetically in an industrial computer. Imaging capability has been proven to offer a novel microwave image quality. Lateral resolution of 2.2 mm and image dynamic range exceeding 30 dB have been demonstrated. Thanks to the relatively fast measurement time, microwave imaging of humans without motion blurring has been successfully achieved.

The accomplished results clearly indicate the advantages of planar multistatic arrays. Technological realization is very challenging, as it requires nevertheless high level of integration on the RF side as well as the digital side. Image reconstruction is a bottleneck in the imaging performance, when it comes to demands for high frame rates. The continuous advances in digital computing, however, can satisfy the future practical requirements. The offered three-dimensional complex-valued images with high dynamic range can be further applied for super-resolution methods or image processing algorithms to extract more features and information about the imaged object. This opens vast opportunities for new applications to effectively utilize this modern form of microwave imaging.

# Scalar Green's Function

The expression of the Green's function $G(\mathbf{r})$ used to solve the Helmholtz's equation [57] for a general scalar field in a three-dimensional space

$$\nabla^2 G(\mathbf{r}) + k^2 G(\mathbf{r}) = -\delta(\mathbf{r}) \tag{A.1}$$

is given next. The $\delta(\mathbf{r})$ is the three-dimensional Dirac delta function; the $\mathbf{r}$ denotes a position vector in the three-dimensional space, such that $\mathbf{r} = r\hat{\mathbf{r}}$ for $\hat{\mathbf{r}}$ a unit vector pointing away from the origin. This development is well-known in literature for solving the scalar wave problems, e.g., in electromagnetics, acoustic, and quantum mechanics. The solution in space domain will be detailed here for the sake of completeness.

An initial expression for a postulated Green's function is given as

$$G(\mathbf{r}) = A \underbrace{\frac{e^{-jkr}}{r}}_{\text{outward}} + B \underbrace{\frac{e^{+jkr}}{r}}_{\text{inward}} , \tag{A.2}$$

where $A$ and $B$ are unknown constants. This describes a superposition of two possible solutions, an outward propagating spherical wave from the origin and an inward absorbed spherical wave in the origin. For the underlying electromagnetic problem, it is physically convenient to keep the first term and drop the second one, where the understanding of any radiation problem indicates an outward radiated wave out of the source.

In order to prove the validity of the postulated expression, the wave equation is evaluated apart of the origin to examine the condition

$$\left(\nabla^2 G(\mathbf{r}) + k^2 G(\mathbf{r})\right)\Big|_{|\mathbf{r}|>0} = 0. \tag{A.3}$$

For this, the Laplacian of $G$ is to be found. Considering that $G$ is of spherical symmetry with respect to $r$, the following thus applies

$$
\begin{aligned}
\nabla^2 G(\mathbf{r}) &= \nabla \cdot \nabla G(\mathbf{r}) \\
&= \nabla \cdot \left( \frac{\partial}{\partial r} G(\mathbf{r})\, \hat{\mathbf{r}} \right) \\
&= \nabla \cdot \left( \frac{\partial}{\partial r} \left( A \frac{e^{-jkr}}{r} \right) \hat{\mathbf{r}} \right) \\
&= \nabla \cdot \left( \left( -A \frac{e^{-jkr}}{r^2} - jkA \frac{e^{-jkr}}{r} \right) \hat{\mathbf{r}} \right) \\
&= \frac{1}{r^2} \frac{\partial}{\partial r}\, r^2 \left( -A \frac{e^{-jkr}}{r^2} - jkA \frac{e^{-jkr}}{r} \right) \\
&= -k^2\, A \frac{e^{-jkr}}{r} \\
&= -k^2\, G(\mathbf{r}) \qquad \text{(A.4)}
\end{aligned}
$$

This hence proves the postulated expression of the Green's function.

In order to calculate the constant $A$, a volume integration on the two hand sides of eq. A.1 is considered for an arbitrary value of $r \neq 0$. Care must be taken in evaluating the volume integration of the Laplacian of $G(\mathbf{r})$, due to the function singularity at the origin. The gradient of the function tends to an infinite value there, and hence the diversion operator cannot be directly applied to the gradient result. Therefore, the volume integration is replaced by a surface integration over a sphere centered at the origin by using Gauss' theorem, which for a vector field $\boldsymbol{\Psi}$ states that

$$
\iiint_V \nabla \cdot \boldsymbol{\Psi}\, dv = \oiint_S \boldsymbol{\Psi} \cdot d\mathbf{s}\,, \qquad \text{(A.5)}
$$

where $d\mathbf{s}$ is a vector pointing out of the surface $S$, which encloses the volume $V$. For the proposed sphere, the $d\mathbf{s}$ and $\hat{\mathbf{r}}$ are aligned, therefore $\hat{\mathbf{r}} \cdot d\mathbf{s} = ds$. Applying the volume integration on eq. A.1 gives

$$
\begin{aligned}
\iiint_V \nabla^2 G(\mathbf{r})\, dv + \iiint_V k^2 G(\mathbf{r}) dv &= -\iiint_V \delta(\mathbf{r}) dv \\
\oiint_S [\nabla G(\mathbf{r})] \cdot d\mathbf{s} + \iiint_V k^2 G(\mathbf{r}) dv &= -1 \qquad \text{(A.6)}
\end{aligned}
$$

The first integral on the left hand side of eq. A.6 is evaluated as follows

$$
\begin{aligned}
\oiint_S [\nabla G(\mathbf{r})] \cdot d\mathbf{s} &= \oiint_S \left[\nabla\left(A\frac{e^{-jkr}}{r}\right)\right] \cdot d\mathbf{s} \\
&= \oiint_S \left[\frac{\partial}{\partial r}\left(A\frac{e^{-jkr}}{r}\right)\right] \hat{\mathbf{r}} \cdot d\mathbf{s} \\
&= \left[\frac{\partial}{\partial r}\left(A\frac{e^{-jkr}}{r}\right)\right] \oiint_S ds \\
&= \left(-A\frac{e^{-jkr}}{r^2} - jkA\frac{e^{-jkr}}{r}\right) \cdot 4\pi r^2 \\
&= -4\pi A\left(e^{-jkr} + jkr\, e^{-jkr}\right) \qquad \text{(A.7)}
\end{aligned}
$$

The second integral on the left hand side of eq. A.6 is evaluated as follows

$$
\begin{aligned}
\iiint_V k^2 G(\mathbf{r}) dv &= \int_0^r k^2 \left(A\frac{e^{-jkr}}{r}\right) \cdot 4\pi r^2 dr \\
&= 4\pi A \int_0^r k^2 r\, e^{-jkr} dr \\
&= 4\pi A \left[jkr\, e^{-jkr} + e^{-jkr}\right]_0^r \\
&= 4\pi A \left[jkr\, e^{-jkr} + e^{-jkr} - 1\right] \qquad \text{(A.8)}
\end{aligned}
$$

Now, by substituting the results of eq. A.7 and eq. A.8 into eq. A.6, it is evident that $A = 1/4\pi$. Accordingly, the Green's function describing an outward propagating spherical scalar wave excited by a $-\delta(\mathbf{r})$ is given by

$$
G(\mathbf{r}) = \frac{e^{-jkr}}{4\pi r}\,. \qquad \text{(A.9)}
$$

By shifting the coordinates in eq. A.1 by and arbitrary vector $\mathbf{r}'$, then the Green's function for an excitation of a $-\delta(\mathbf{r} - \mathbf{r}')$ is given by

$$
G(\mathbf{r}) = \frac{e^{-jk|\mathbf{r}-\mathbf{r}'|}}{4\pi|\mathbf{r} - \mathbf{r}'|}\,. \qquad \text{(A.10)}
$$

Appendix B

# Fourier Transform Definitions

---

The expressions for Fourier transforms vary among the scientific fields, e.g., optics and electromagnetics. However yielding the same meaning, the descriptions could be different. In the following formulations, the definitions for the Fourier transforms $\mathcal{F}$ and their inversions $\mathcal{F}^{-1}$ in one-, two-, and three-dimensional cases are listed. For a function $f$ of $n$ variables, i.e., $\alpha_1$ to $\alpha_n$, its Fourier transform in $n$-dimensions results in a spectrum $F$ with also $n$ variables, i.e., $k_{\alpha_1}$ to $k_{\alpha_n}$, respectively. The function $f$ is generally complex, and assumed to be integrable.

For the one-dimensional case:

$$F(k_\alpha) = \mathcal{F}_{\text{1D}}\left\{f(\alpha)\right\} = \int\limits_{\forall\alpha} f \cdot e^{-jk_\alpha \alpha}\, d\alpha \tag{B.1}$$

$$f(\alpha) = \mathcal{F}_{\text{1D}}^{-1}\left\{F(k_\alpha)\right\} = \frac{1}{2\pi}\int\limits_{\forall k_\alpha} F \cdot e^{+jk_\alpha \alpha}\, dk_\alpha \tag{B.2}$$

For the two-dimensional case:

$$\begin{aligned} F(k_{\alpha_1}, k_{\alpha_2}) &= \mathcal{F}_{\text{2D}}\left\{f(\alpha_1, \alpha_2)\right\} \\ &= \int\limits_{\forall\alpha_1}\int\limits_{\forall\alpha_2} f \cdot e^{-j(k_{\alpha_1}\alpha_1 + k_{\alpha_2}\alpha_2)}\, d\alpha_1\, d\alpha_2 \end{aligned} \tag{B.3}$$

$$\begin{aligned} f(\alpha_1, \alpha_2) &= \mathcal{F}_{\text{2D}}^{-1}\left\{F(k_{\alpha_1}, k_{\alpha_2})\right\} \\ &= \frac{1}{(2\pi)^2}\int\limits_{\forall k_{\alpha_1}}\int\limits_{\forall k_{\alpha_2}} F \cdot e^{+j(k_{\alpha_1}\alpha_1 + k_{\alpha_2}\alpha_2)}\, dk_{\alpha_1}\, dk_{\alpha_2} \end{aligned} \tag{B.4}$$

For the three-dimensional case:

$$F(k_{\alpha_1}, k_{\alpha_2}, k_{\alpha_3}) = \mathcal{F}_{3D}\{f(\alpha_1, \alpha_2, \alpha_3)\} \tag{B.5}$$
$$= \int\limits_{\forall\alpha_1}\int\limits_{\forall\alpha_2}\int\limits_{\forall\alpha_3} f \cdot e^{-j(k_{\alpha_1}\alpha_1 + k_{\alpha_2}\alpha_2 + k_{\alpha_3}\alpha_3)}\, d\alpha_1\, d\alpha_2\, d\alpha_3$$

$$f(\alpha_1, \alpha_2, \alpha_3) = \mathcal{F}_{3D}^{-1}\{F(k_{\alpha_1}, k_{\alpha_2}, k_{\alpha_3})\} \tag{B.6}$$
$$= \frac{1}{(2\pi)^3}\int\limits_{\forall k_{\alpha_1}}\int\limits_{\forall k_{\alpha_2}}\int\limits_{\forall k_{\alpha_3}} F \cdot e^{+j(k_{\alpha_1}\alpha_1 + k_{\alpha_2}\alpha_2 + k_{\alpha_3}\alpha_3)}\, dk_{\alpha_1}\, dk_{\alpha_2}\, dk_{\alpha_3}$$

For an n-dimensional case:

$$F(k_{\alpha_1}, \cdots, k_{\alpha_n}) = \mathcal{F}_{nD}\{f(\alpha_1, \cdots, \alpha_n)\} \tag{B.7}$$
$$= \int\limits_{\forall\alpha_1} \cdots \int\limits_{\forall\alpha_n} f \cdot e^{-j(k_{\alpha_1}\alpha_1 + \cdots\cdots + k_{\alpha_n}\alpha_n)}\, d\alpha_1 \cdots d\alpha_n$$

$$f(\alpha_1, \cdots, \alpha_n) = \mathcal{F}_{nD}^{-1}\{F(k_{\alpha_1}, \cdots, k_{\alpha_n})\} \tag{B.8}$$
$$= \frac{1}{(2\pi)^n}\int\limits_{\forall k_{\alpha_1}} \cdots \int\limits_{\forall k_{\alpha_n}} F \cdot e^{+j(k_{\alpha_1}\alpha_1 + \cdots\cdots + k_{\alpha_n}\alpha_n)}\, dk_{\alpha_1} \cdots dk_{\alpha_n}$$

Appendix C

# Fourier Transform over 2D Aperture

The calculations of the two-dimensional Fourier transform of the Tx and Rx terms given in sec. 2.6.2 are described next. These transformations are essential for the spatial frequency domain representation of the measured signals. As both terms have an identical form, the solution steps are shown for the Tx term and then the solution for the Rx term is similarly written.

The Tx term, named here $s^T$, is given as

$$\begin{aligned} s^T &= \exp\left(-jk\sqrt{(x_t-x)^2+(y_t-y)^2+(z_a-z)^2}\right) \\ &= \exp\left(-jkR_t\right), \end{aligned} \tag{C.1}$$

where $x_t$ and $y_t$ describe the lateral distribution of the transmitting planar aperture $z = z_a$ and $R_t$ physically describes the geometrical distance between the transmitter position $(x_t, y_t, z_a)$ and the object position $(x, y, z)$. Similarly, the Rx term, named here $s^R$, is given as

$$\begin{aligned} s^R &= \exp\left(-jk\sqrt{(x_r-x)^2+(y_r-y)^2+(z_a-z)^2}\right) \\ &= \exp\left(-jkR_r\right), \end{aligned} \tag{C.2}$$

where $x_r$ and $y_r$ describe the lateral distribution of the receiving planar aperture $z = z_a$ and $R_r$ physically describes the geometrical distance between the receiver position $(x_r, y_r, z_a)$ and the object position $(x, y, z)$.

Considering the Tx term, it is thus required to find the 2D Fourier transform of $s^T$ with respect to $x_t$ and $y_t$, namely

$$S^T(k_{x_t}, k_{y_t}) = \mathcal{F}_{\text{2D}}\left\{s^T\right\}. \tag{C.3}$$

The Fourier transform is expressed as

$$\mathcal{F}_{\text{2D}}\left\{s^T\right\} = \int\limits_{x_t}\int\limits_{y_t} s^T \cdot \exp(-j(k_{x_t}x_t + k_{y_t}y_t))\, dx_t\, dy_t \tag{C.4}$$

$$= \int\limits_{x_t}\int\limits_{y_t} \exp(+j\underbrace{(-kR_t - k_{x_t}x_t - k_{y_t}y_t)}_{\text{phase } \Phi})\, dx_t\, dy_t \tag{C.5}$$

This is an integration over an oscillatory complex function, which can be solved using asymptotic integration methods. A commonly used method for this type of integration is the method of stationary phase, discussed in detail in [119] and [57]. According to it, the integration value is given by the sum of the function values at the positions where the phase is stationary, i.e., of zero first derivative. This is physically understood by considering that the main contribution of the integrated function comes from the points where the values add constructively and hence the local phase is stationary; whereas the other contributions will add destructively due to the phase oscillating behavior of the function. The magnitude of the result depends on the curvature of the function at the neighborhood of the stationary points, which can be calculated from the higher derivatives of the integrated function. However in the presented application, the magnitude does not play an important role in the imaging process and hence is ignored. Therefore, $S^T$ is considered mathematically to be normalized.

As it is shown next, the phase $\Phi$ has one stationary point, named to be $(x_t^0, y_t^0)$. This stationary point satisfies

$$\frac{\partial}{\partial x_t}\Phi(x_t, y_t) = 0 \text{ and} \tag{C.6}$$

$$\frac{\partial}{\partial y_t}\Phi(x_t, y_t) = 0\,, \tag{C.7}$$

for $x_t = x_t^0$ and $y_t = y_t^0$. The solution is accordingly formulated as in [119] by

$$S^T = \frac{2\pi j}{\sqrt{\Phi_{x_t x_t}\Phi_{y_t y_t} - \Phi_{x_t y_t}^2}} \cdot \exp\left(+j\Phi(x_t^0, y_t^0)\right)$$

$$\approx \exp\left(+j\Phi(x_t^0, y_t^0)\right)\,, \tag{C.8}$$

where $\Phi_{x_t x_t}$, $\Phi_{y_t y_t}$, and $\Phi_{x_t y_t}$ denote the double partial differentiations with respect to $x_t$ and $y_t$ and evaluated at $(x_t^0, y_t^0)$.

Applying eq. C.6 and eq. C.7 will, respectively, result in

$$R_t^0 \cdot k_{x_t} = -k(x_t^0 - x) \text{ and} \tag{C.9}$$

$$R_t^0 \cdot k_{y_t} = -k(y_t^0 - y) \ , \tag{C.10}$$

where $R_t^0$ denotes the value of $R_t$ at the stationary point. By first squaring eq. C.9 and eq. C.10, then adding them, it yields

$$(R_t^0)^2 = \frac{k^2 \cdot (z_a - z)^2}{k^2 - k_{x_t}^2 - k_{y_t}^2} \ . \tag{C.11}$$

Considering that:

- $k > 0$
- $z > z_a \Rightarrow (z_a - z) < 0$, hence $\sqrt{(z_a - z)^2} = -(z_a - z)$
- $k^2 > (k_{x_t}^2 + k_{y_t}^2)$ in practice for all $z > z_a$

then eq. C.11 reduces to

$$R_t^0 = \frac{-k \cdot (z_a - z)}{\sqrt{k^2 - k_{x_t}^2 - k_{y_t}^2}} \ . \tag{C.12}$$

By substituting eq. C.9, eq. C.10, and eq. C.12 into eq. C.8 and doing some algebraic arrangements, the final solution becomes

$$S^T = \exp\left(+j\sqrt{k^2 - k_{x_t}^2 - k_{y_t}^2} \cdot (z_a - z) - jk_{x_t}x - jk_{y_t}y\right) . \tag{C.13}$$

Following the same steps for the Rx term $s^R$, its 2D Fourier transform is expressed as

$$S^R = \exp\left(+j\sqrt{k^2 - k_{x_r}^2 - k_{y_r}^2} \cdot (z_a - z) - jk_{x_r}x - jk_{y_r}y\right) . \tag{C.14}$$

# Numerical example

A numerical example for illustration purpose is prepared using the parameters in table C.1. The corresponding phase distribution $\Phi(x_t, y_t)$ is shown in fig. C.1. By substituting the numerical values into eq. C.12, then into eq. C.9 and eq. C.10, the stationary point $(x_t^0, y_t^0)$ is calculated to exist at the location $(-0.0675, 0.0462)$ m. This point is annotated in fig. C.1, thus exactly predicting the position of the stationary point. The oscillatory behavior of the function is also seen in its far neighborhood.

| Parameter | Value |
|---|---|
| Frequency | 75 GHz |
| $(x, y, z)$ | $(0.04, 0.1, 0.05)$ m |
| $z_a$ | $-0.5$ m |
| $k_{x_t}$ | 300 rad/m |
| $k_{y_t}$ | 150 rad/m |

Table C.1: Values used in the numerical example.

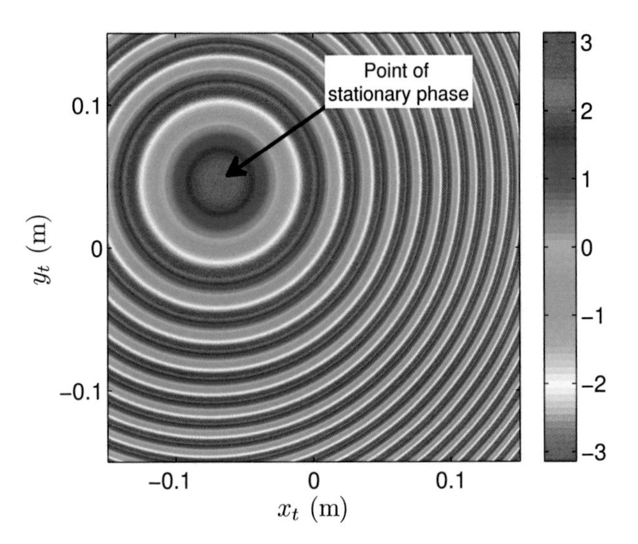


Figure C.1: An illustrating example for the application of the method of stationery phase. The phase $\Phi(x_t, y_t)$ is shown in radians for the values in table C.1.

Appendix D

# Image Gallery

In this appendix, a selection of microwave images is presented with the aim of expanding the understanding of the rich content of microwave images delivered by the introduced imaging method. This understanding is essential for proper interpretation of microwave images, and hence also for efficient utilization of the microwave imaging capabilities. The images are mostly presented with explanation, however, the emphasis is mostly made on the visualization of the image information content rather than the measurement method or the detailed signal processing steps. Examples addressing imaging of humans as well as the reflectivity level, the penetration capability, and the scattering behavior at the millimeter-wave range are presented. In addition, an advanced view for the phase information is presented for demonstration of the rich content of the complex-valued images. Lastly, an example point spread function discussed previously in chapter 2 is three-dimensionally visualized.

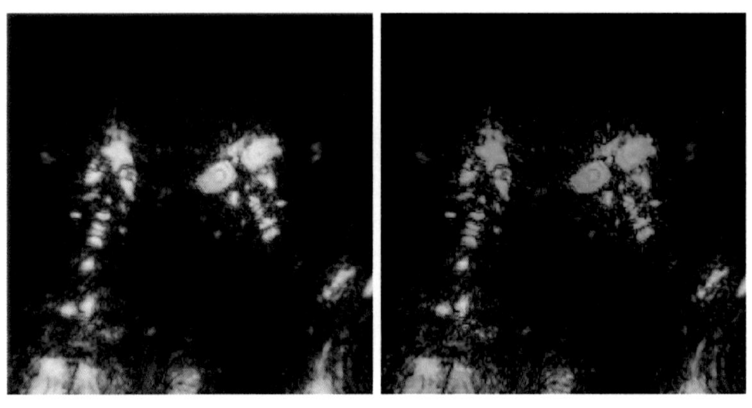

(a) Sepia-toned magnitude image (b) Depth-colored magnitude image

Figure D.1: The very first image of humans taken with the fully electronic demonstrator presented in chapter 5. It is considered the first image of its kind to successfully image humans in E-band without any mechanical scanning and while utilizing active illumination. The image is dated to the 18th of September 2010.

A complete visualization of the reconstructed 3D complex-valued images in a single view is close to impossible even with modern means. Instead, the image must be interpreted in different ways, and hence appropriate visualizations are prepared for viewing its content. Each voxel in the 3D image assembles a magnitude and a phase value. The magnitude value corresponds directly to the reflectivity of the object at the voxel position. The phase value corresponds, however, to the phase response of the object and simultaneously to the electrical length given by the distance between the voxel center and the actual position of the object. While imaging with a signal bandwidth, this electrical length is approximately determined by the center frequency.

# Magnitude information

The magnitude information of the complex-valued image can be visualized in 3D. Considering the 3D magnitude image, its content can be understood as a distribution of voxels on a 3D grid with their magnitudes mapped to a transparency level, e.g., alpha value in the terminology of computer graphical visualizations. Another treatment can be made using a point cloud model, in which the data is assumed to be scattered in space and is not necessarily bound to a 3D grid. Additionally, the 3D magnitude image can be projected along any direction. However being the simplest visualization, most of the relevant information is usually available in the projected image along range direction, an example is seen in fig. D.1. Optionally, the depth information at each projected voxel might be used to color the image. This thus results in a combined view for the reflectivity of the voxel and its range position.

The figures D.2 and D.3 illustrate the issue of illumination on complex 3D objects. The image is the one presented in sec. 2.9.3. The magnitude image is rendered in 3D and is mapped to a priori known 3D surface model of the mannequin. This surface model is visualized in blue color, and is then combined with the microwave image and the imaging array itself in a full visualization. The visible parts of the surface in the microwave image clearly correspond to the relations discussed in chapter 3. The left arm, for instance, appears illuminated in the microwave image to some extent where the normal vectors of the surface still pass through the illumination effective aperture. Its lower part, however, appears dark because of its orientation. The same argumentation applies for the shoulders. It is also worth observing that the scattering of the pullover can still be seen in the foreground of the mannequin. In other cases, the visualization of the 3D magnitude image can then reveal the layered structure of the imaged object, e.g., the case with the pullover and the underneath surface.

Figure D.2: Photograph of the dressed mannequin. Its surface was metalized in order to produce similar reflectivity as the human skin.

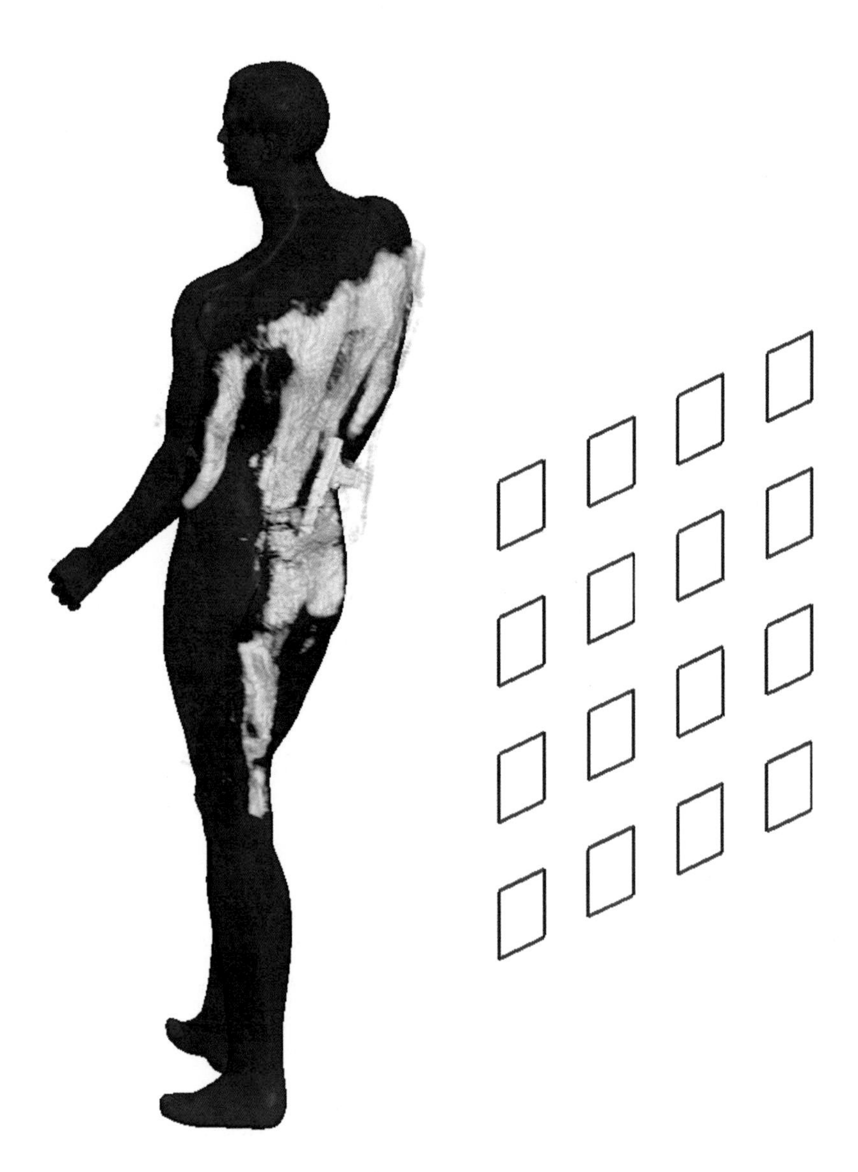

Figure D.3: On the left side, 3D rendering of a complete surface model of the mannequin overlapped with a rendering of the magnitude of the 3D microwave image. On the right side, the imaging array of $1m^2$ aperture is illustrated to scale. A slant view is used to illustrate the 3D arrangement.

# Phase information

Care must be taken when evaluating the phase information, as the phases should be considered accurate only where significant level of magnitude is present. Otherwise, strong disturbance due to noise and possible artifacts can falsify the phase readings. Moreover, phase information is better not viewed directly in 3D, instead, it should be processed to extract certain relations corresponding to geometrical features or electrical properties. For this, it is helpful to consider that the phase value at a voxel is approximately determined by $(\phi_{\text{obj}} - 4\pi\,\delta d \cdot f^{\text{mid}}/c_0)$, where $\phi_{\text{obj}}$, $\delta d$, and $f^{\text{mid}}$ denote the phase response of the object, the distance between the voxel center and the object measured along the $z$-direction, and the center frequency, respectively. An example of the utilization of the phase information for the determination of surface normals is illustrated next.

Fig. D.4 shows a colored hemisphere used as a legend for visualizing the surface local normal vectors. Each vector direction is mapped to a unique color. This is achieved by using the slant angle $\Theta$ to modulate the brightness level, whereas the rotation angle $\Phi$ determines the color tone. In this specific definition, the reference axes can be chosen arbitrarily without having influence on the continuity of the color mapping. For surfaces parallel to the imaging array, $\Theta$ is equal to zero. Fig. D.5 now demonstrates the usage the phase values to extract more information regarding the geometry of the object's surface.

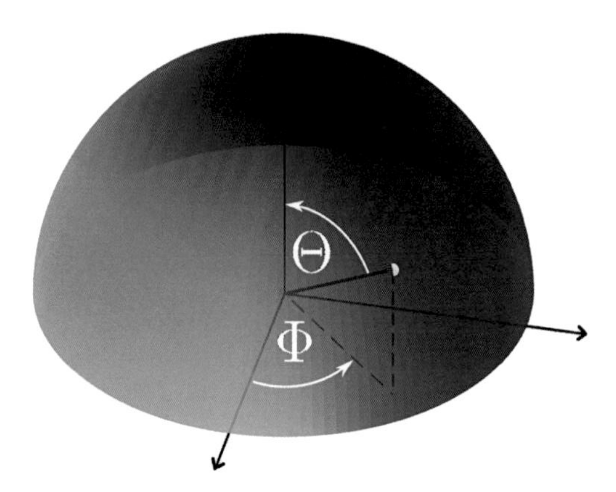


Figure D.4: A 3D view of the legend used to display the direction of the surface local normal vector dependent on the slant angle $\Theta$ and the orientation angle $\Phi$. In a 2D projection view, the radius thus corresponds to $\sin\Theta$, which is chosen to modulate the brightness level from zero to one. The value of $\Phi$, however, modulates the color irrelevant of the brightness. The rotation color is discontinuous due to the used HSV color scale [83].

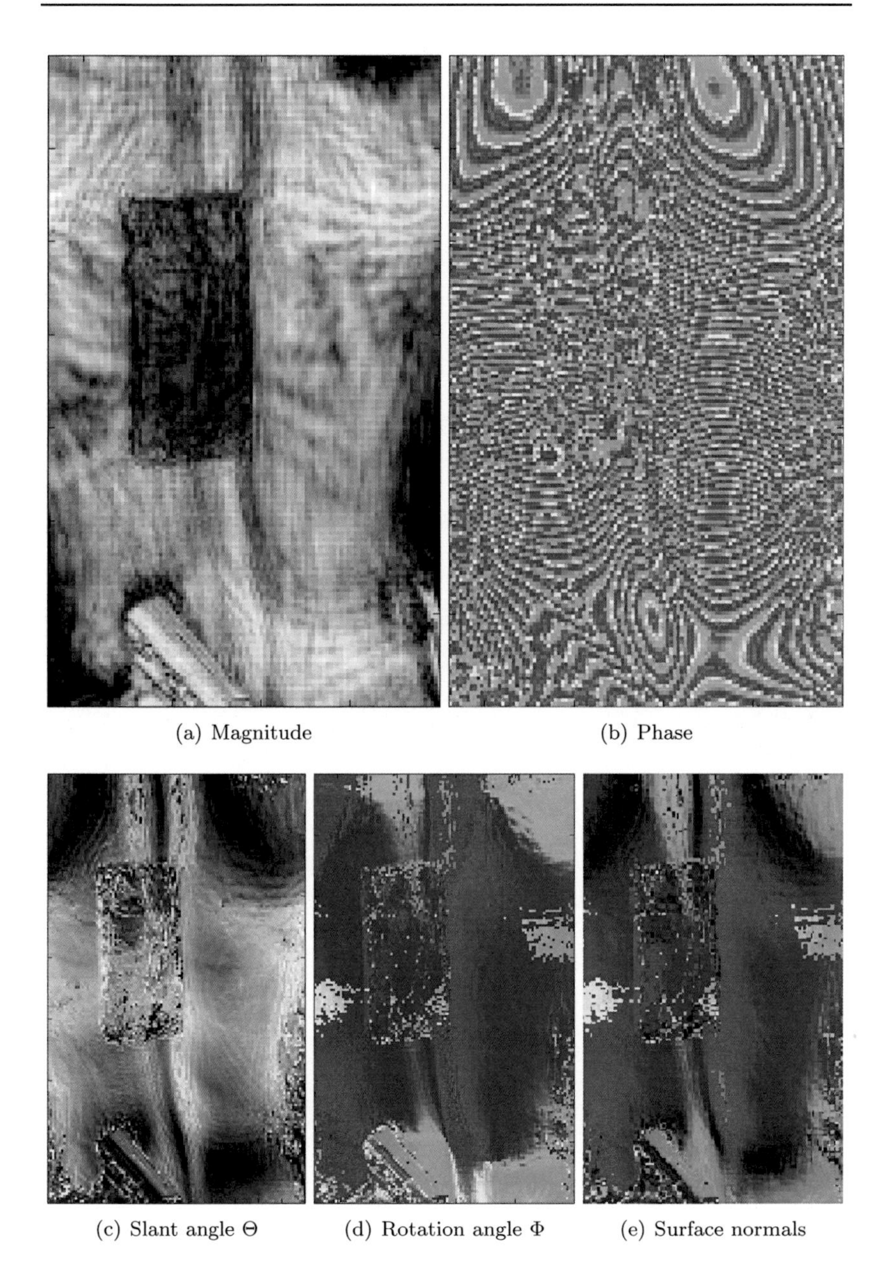

(a) Magnitude (b) Phase

(c) Slant angle $\Theta$ (d) Rotation angle $\Phi$ (e) Surface normals

Figure D.5: Example of the usage of phase information. The phases are considered at the positions of maximum magnitude. Although it may seem chaotic at the first glance, phase gradients offer accurate information about surface normals, leading to a detailed view of object's geometry.

## Reflectivity

Materials have varying reflectivity depending on the imaging frequency. It is therefore interesting to determine the reflectivity of objects and materials considered in the imaging applications utilizing millimeter-waves. Two examples are shown next, using a frequency range of 70 to 80 GHz. The first is shown in fig. D.6. A hand is imaged while a metal strip is prepared on the index finger. Care is here taken to keep similar surface geometry of the strip and the skin itself, thus avoiding change of reflectivity due to illumination issues. Fig. D.6(b) gives an impression about the reflectivity. The image is normalized to the metal reflections and presented in linear scale. The skin here shows around 60% reflectivity compared to metal. This is equivalent to −4.5 dB.

The second example compares the reflectivity of ceramic and metal knifes. Being a significant threat to passenger security, detection of ceramic knifes at airport security checkpoints with millimeter-wave imagers is highly demanded by aviation security organizations. Here, a direct comparison is conducted by imaging both knifes, i.e., a metal and a ceramic one, simultaneously. Fig. D.7 presents the result in a 30 dB logarithmic view. The ceramic knife is clearly visible and shows a reflectivity level of approximately −3 dB compared to metal. This corresponds to almost 70% reflectivity. Nicely, the extension of the metal bevel inside the grip of the metal knife is easily identified in the microwave image.

## Scattering

The influence of the two main scattering phenomena, namely the specular and diffuse ones, on the illumination quality in microwave images have been comprehensively discussed in chapter 3. An elementary example in fig. D.8 is now shown to demonstrate these two fundamental behaviors. A flat and thin metal sheet is imaged twice with different orientations. The first time, fig. D.8(b), corresponds to the case of specular reflection, for which the metal sheet is oriented parallel to the aperture and centered to it. This ensures that the normal vectors at all surface portions to cross the illumination effective aperture of the imaging array. Furthermore, the sheet is tilted during the second time such that all normal vectors cannot cross the illumination effective aperture as before. Fig. D.8(c) presents the image in this case. While the inner surface of the object becomes completely dark, edge scattering causes the boundaries of the sheet to be clearly visible. Thanks to the high processing gain made by the reconstruction process, the diffuse scattering at the edges reaches a significant level of brightness making it visible in the focused image.

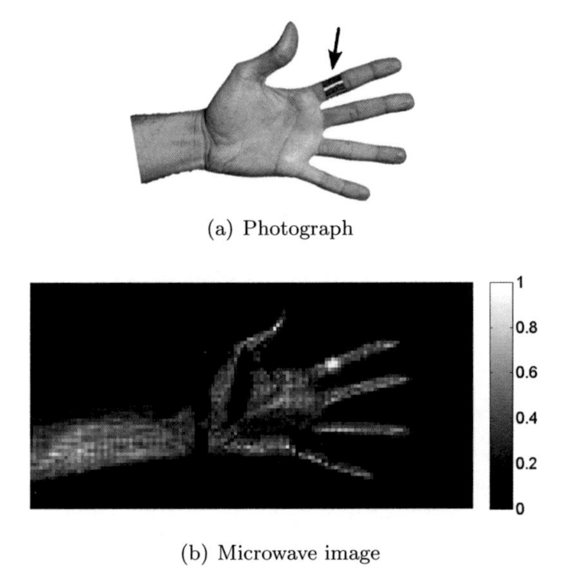


(a) Photograph

(b) Microwave image

Figure D.6: Human skin shows around 60% reflectivity compared to a metal strip. Image is in linear scale normalized to the metal reflections.

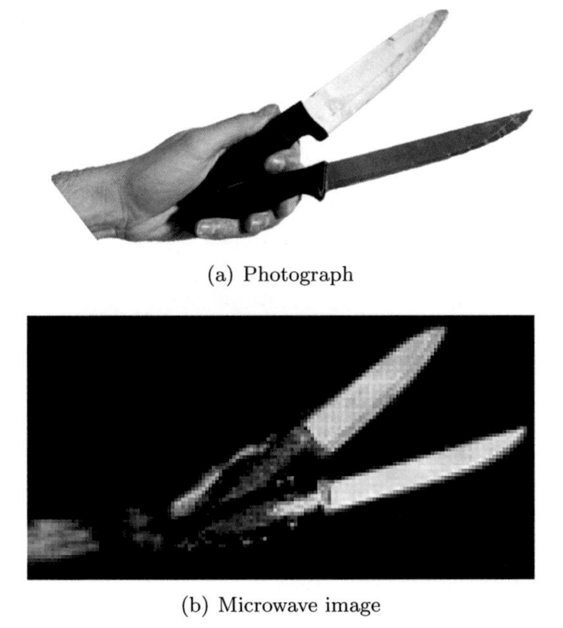

(a) Photograph

(b) Microwave image

Figure D.7: Comparison between metal and ceramic knifes in a 30 dB view. The ceramic knife shows approximately 3 dB less reflections.

(a) Photograph

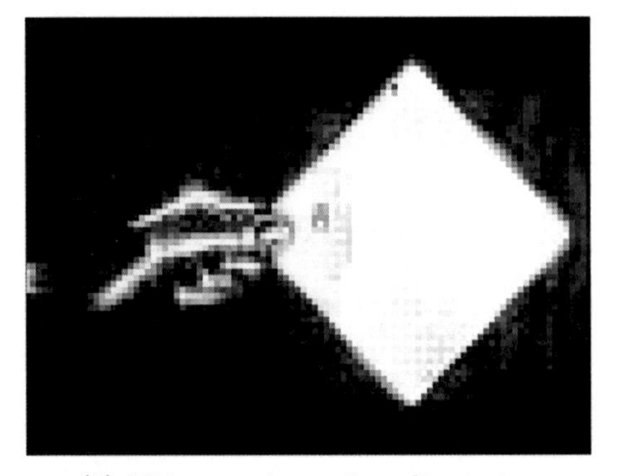

(b) Microwave image in a direct view

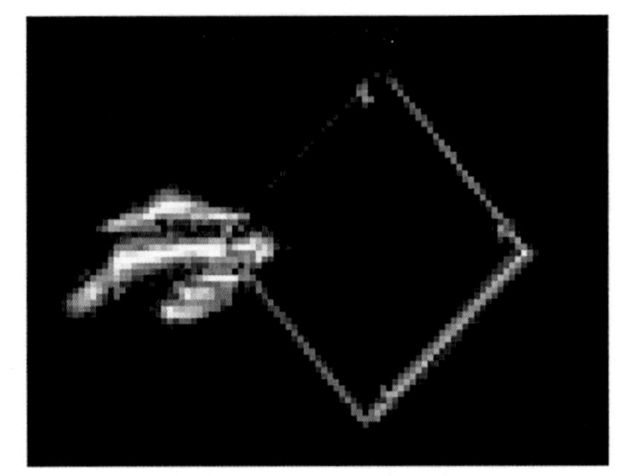

(c) Microwave image in a tilted view

Figure D.8: Demonstration of the two main scattering behaviors and their influence on the imaging result. Images are made using a frequency range of 70 to 80 GHz. A flat and thin metal sheet is imaged twice in two different orientations in order to illustrate specular reflection case on its surface and diffuse reflection case at its edges.

# Penetration

A major advantage of microwave imaging is its capability to penetrate many optically opaque surfaces. In the millimeter-wave range, i.e., here 70 to 80 GHz, many non-metallic objects from everyday use become transparent. The following examples demonstrate this feature and are selected to broadly address various interests for applications. Imaging through cloths, for instance, is illustrated by fig. D.9, where a part of a person's hand is optically hidden in a trouser pocket. The microwave image clearly reveals the hand with a very little distortion caused by the cloth. Penetration through bulk dielectric materials is also achievable, especially when losses are not high. An example is shown in fig. D.10 for a thick Polytetrafluoroethylene block, e.g., widely known as Teflon and its permittivity is approximately equal to 2.1 [120]. The block is prepared partially with metal on its background in order to prove its transparency. The brightness of the metal surface imaged through the thick block is evident. This is very beneficial for applications targeting material analysis, either for material characterization or quality control purposes.

Penetration through usual objects is also examined. Imaging through a closed box and through a package is shown next. Fig. D.11 illustrates an imaging result for a plastic box including a bottle opener made of metal. The location and the shape of the hidden object is easily identified. Imaging through cartoon packages is demonstrated by fig. D.12. The package is imaged while being closed from the bottom side. The microwave image is prepared to visualize the inside volume of the package after geometrical filtering of the cartoon itself. With a resolution of a few millimeters, the contents of the package can be easily interpreted.

Personnel screening has become recommended for enhancing the security checkpoints at airports and critical infrastructure building. The millimeter-waves are very suitable for imaging humans [90, 92]. The good reflectivity of the human skin encourages the utilization of this imaging method to identify concealed objects and contraband weapons. Besides the detection of metallic objects, e.g., guns, detection of dielectric objects is required. Among them, ceramic knifes, plastic, and liquid explosives are potential threats to the security of passengers. As addressed above, ceramic knifes are easily visible in the investigated frequency range of 70 to 80 GHz. In the following example, an image of a human taken from a rear view using the system in [121] is presented. The three different methods for the visualization of magnitude information of the 3D image are shown in fig. D.13. Two concealed objects are attached to the imaged person, i.e., a liquid bag and a modeling clay. Both are considered a good surrogate for real explosives, thus showing similar reflectivity [51]. Detailed views for the revealed concealed objects are illustrated in fig D.14.

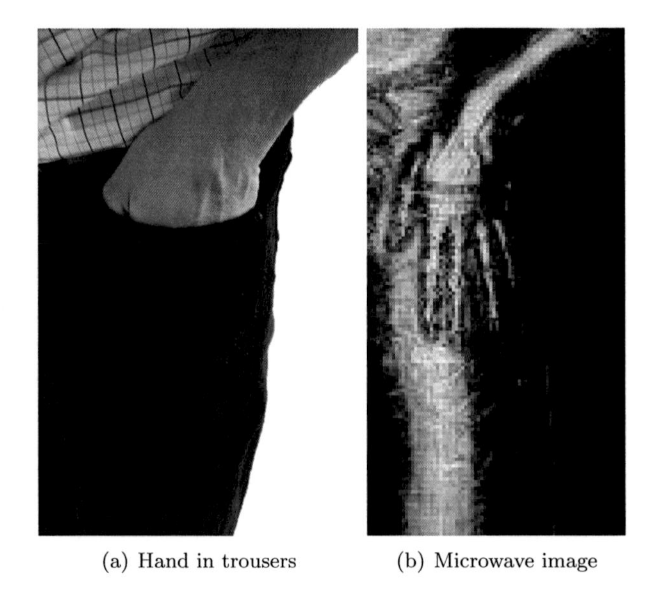

(a) Hand in trousers (b) Microwave image

Figure D.9: Visibility through clothing. Illumination comes from the left side of the image.

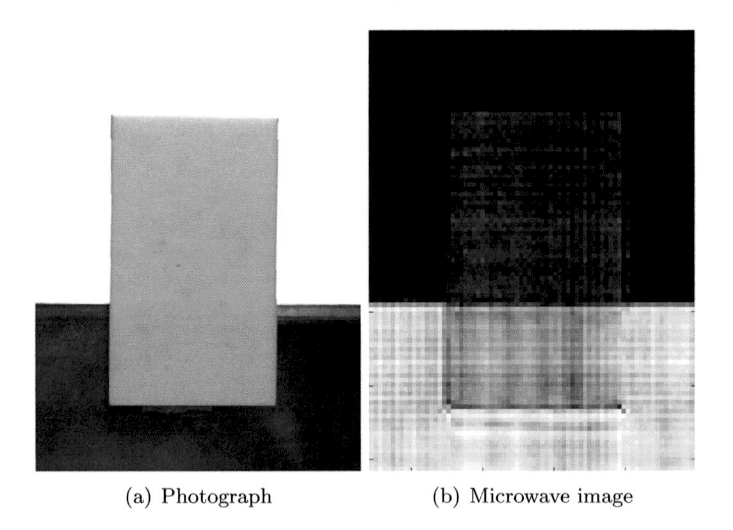

(a) Photograph (b) Microwave image

Figure D.10: Penetration through a Polytetrafluoroethylene (Teflon) block of 11 mm thickness and an extension of 210 mm times 120 mm.

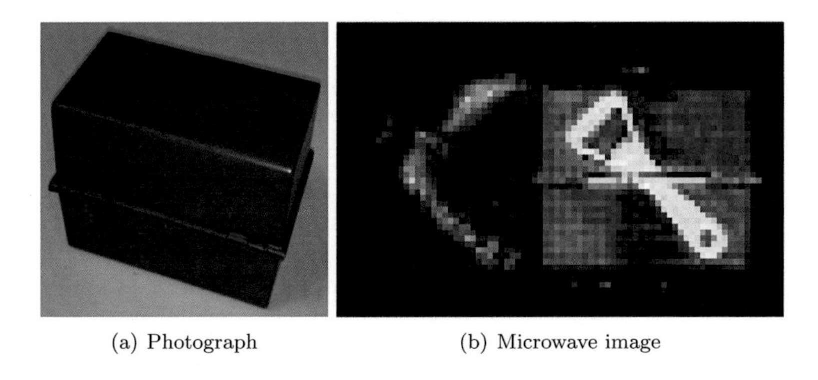

(a) Photograph (b) Microwave image

Figure D.11: Visibility through a plastic box. A hidden bottle opener is clearly revealed in microwave image.

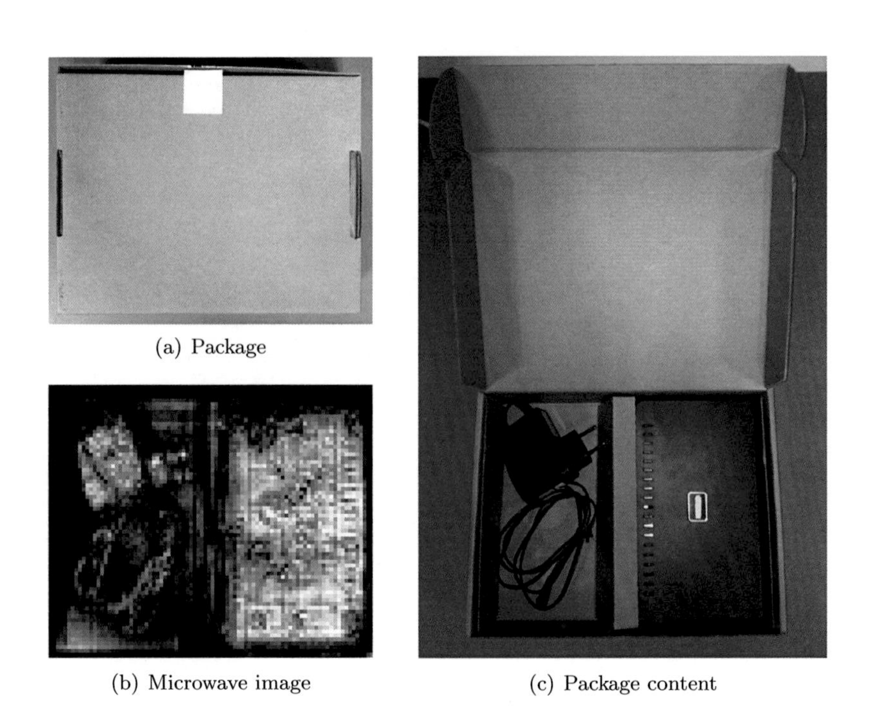

(a) Package

(b) Microwave image (c) Package content

Figure D.12: A view through the inside volume of a closed package using microwave imaging.

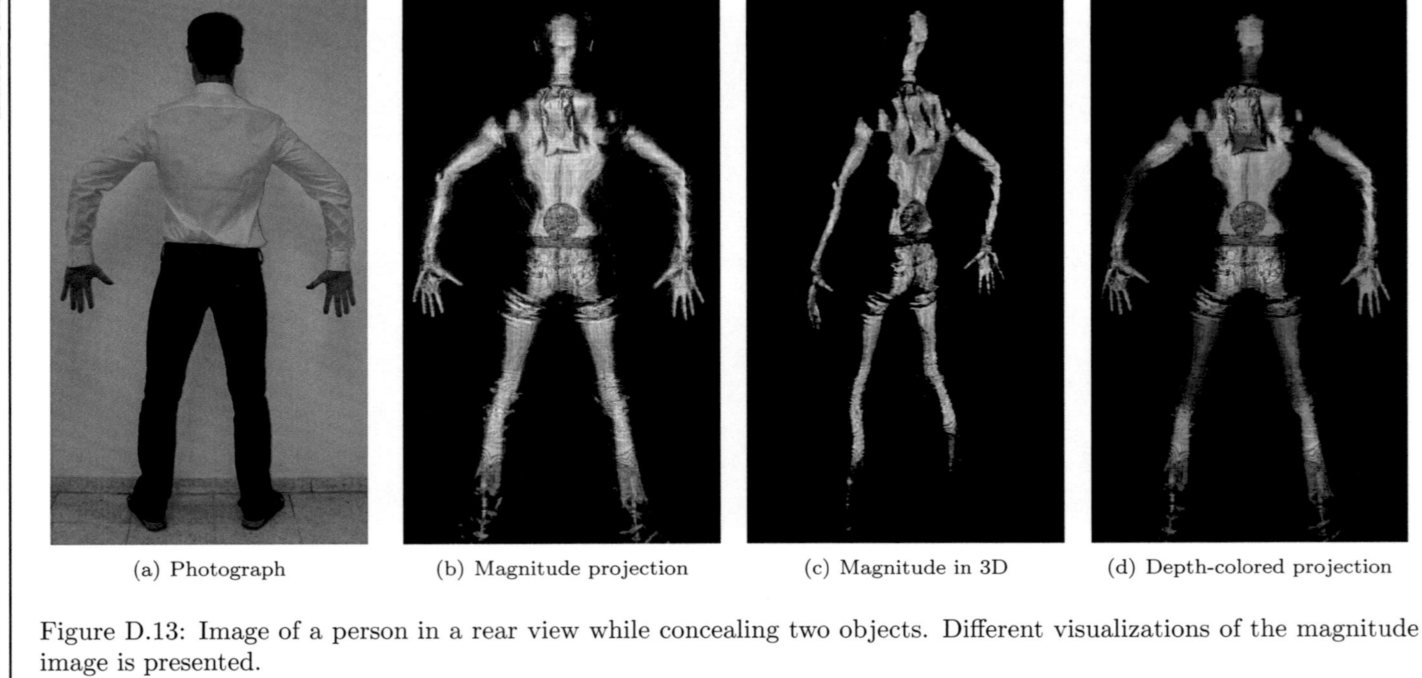

(a) Photograph

(b) Magnitude projection

(c) Magnitude in 3D

(d) Depth-colored projection

Figure D.13: Image of a person in a rear view while concealing two objects. Different visualizations of the magnitude image is presented.

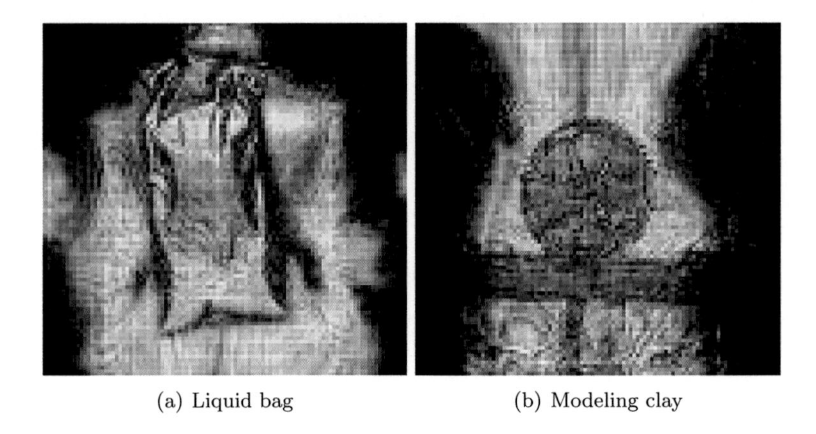

(a) Liquid bag (b) Modeling clay

Figure D.14: Detailed views for the concealed objects seen in fig. D.13.

## Point spread function

This section is dedicated for visualizing[§] the PSF discussed in sec. 2.9.2. However being a specific case, it is meant to be an example to deepen the understanding of its three-dimensional construction, which is approximately also valid for other planar multistatic array. The case for a center position is thus visualized in fig. D.15. The shape of the focused spot along with the surrounding close side-lobes are rendered to scale in fig. D.15(a). Actually, the side-lobes are below −20 dB, but the visualization emphasizes the color tone to make them more visible. When considering other positions instead of the center one, the structure of the PSF would get tilted in 3D space followed by the surrounding side-lobes. Furthermore, the main spot keeps almost oriented towards the center of the aperture.

The corresponding 3D support in the K-space is also visualized in fig. D.15(b) and fig. D.15(c) for a view from the inside of the 2k-hemisphere and from the outside of it, respectively. The visualized support is identical to the one presented previously in fig. 2.23. Multistatic coverage results in a characteristic shape of the support in K-space; this is identified by the tip seen in the inside part of the support. As outlined in fig. 2.7, this characteristic shape leads to a significant enhancement in the range resolution due to the increased support along $k_z$-direction. Moreover, the outer part of the support follows exactly the surface of the $2k^{\mathrm{max}}$-hemisphere, namely for the 80 GHz signal.

[§]The rendering of the three-dimensional data is credited to Dr.-Ing. Cyrille Maire.

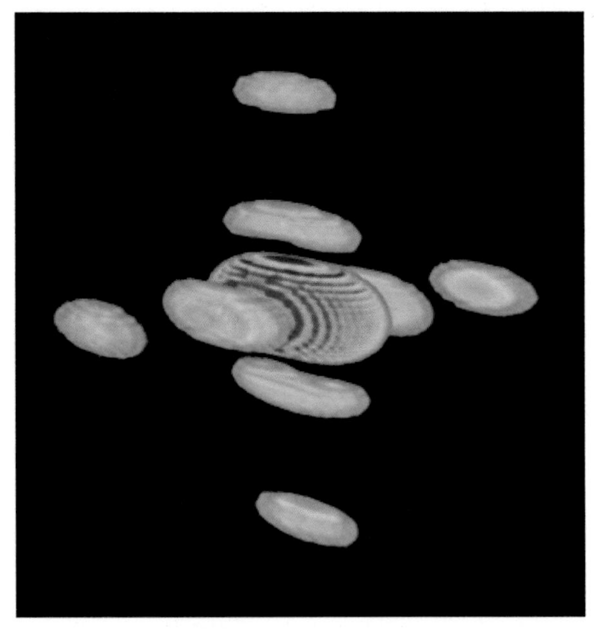

(a) PSF at a center position

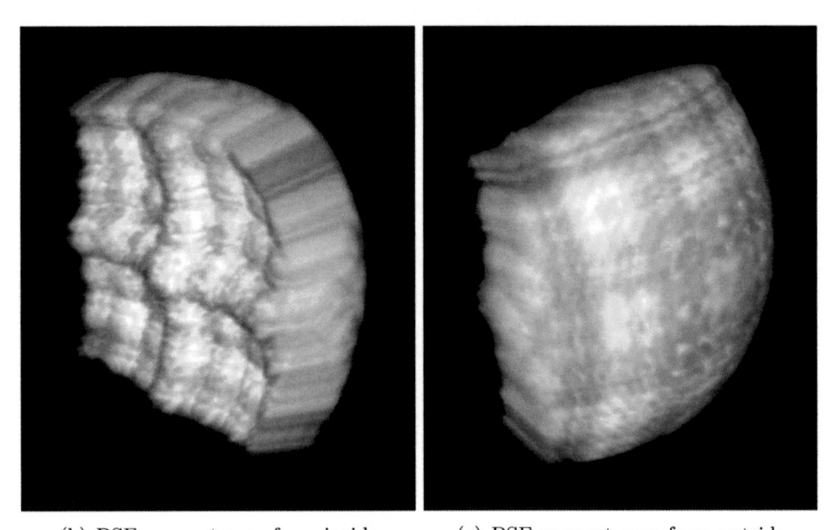

(b) PSF support seen from inside

(c) PSF support seen from outside

Figure D.15: Rendering of the PSF presented in sec. 2.9.2 in 3D along with its support in K-space. The characteristic shape of the multistatic coverage is identified by the small tip in the middle of the support volume.

# Symbols and Acronyms

| Symbol | Description |
|---|---|
| $\mathscr{E}$ | Electric field |
| $\mathbf{E}$ | Electric field phasor |
| $\widehat{E}_{ant}$ | Polarization unit vector of antenna |
| $\mathscr{D}$ | Electric flux density |
| $\mathbf{D}$ | Electric flux density phasor |
| $\mathscr{H}$ | Magnetic field |
| $\mathbf{H}$ | Magnetic field phasor |
| $\mathscr{B}$ | Magnetic flux density |
| $\mathbf{B}$ | Magnetic flux density phasor |
| $j$ | The complex number $\sqrt{-1}$ |
| $\mathscr{J}$ | Electric current density |
| $\rho$ | In sec. 2.1: Electric charge density |
| $\rho$ | In sec. 4.4: Distance between specular point and center of curvature of scattered signal |
| $\boldsymbol{\rho}$ | Electric charge density phasor |
| $\mathscr{Q}$ | Electric charge |
| $c$ | Speed of light |
| $c_0$ | Speed of light in free space |
| $t$ | Time |
| $\omega$ | Angular frequency |
| $f$ | Temporal frequency |
| $f^{\max}$ | Maximum frequency of operation |
| $f^{\min}$ | Minimum frequency of operation |
| $\Delta f$ | Signal bandwidth |

| Symbol | Description |
|---|---|
| $\varepsilon$ | Permittivity |
| $\varepsilon_0$ | Vacuum permittivity |
| $\varepsilon_r$ | Relative permittivity |
| $\varepsilon_r^c$ | Complex relative permittivity |
| $\mu$ | Permeability |
| $\mu_0$ | Vacuum permeability |
| $\sigma$ | Conductivity |
| $\lambda$ | Wavelength |
| $\lambda_0$ | Wavelength in free space |
| $k$ | Wavenumber |
| $k_0$ | Wavenumber in free space |
| $O$ | Object function in space domain |
| $\mathcal{O}$ | Object function in spatial frequency domain |
| $\widehat{O}$ | Reconstruction of $O$ in space domain |
| $s$ | Scattered field in space domain |
| $S$ | Scattered field in spatial frequency domain |
| $\mathbf{s}$ | Position vector of specular point on sphere |
| $\mathbf{r}$ | Three-dimensional position vector |
| $r$ | Length of the position vector $\mathbf{r}$ |
| $\mathbf{r}_v$ | Position vector of target voxel |
| $\mathbf{r}_t$ | Position vector of Tx |
| $\mathbf{r}_r$ | Position vector of Rx |
| $U$ | Scalar field |
| $U^i$ | Incident field |
| $U^s$ | Scattered field |
| $G$ | Scalar Green's function |
| $\delta$ | Three-dimensional Dirac delta function |
| $\delta_x$ | Spatial resolution in $x$-direction |
| $\delta_y$ | Spatial resolution in $y$-direction |
| $\delta_z$ | Spatial resolution in $z$-direction |
| $D$ | Side length of square aperture |
| $L$ | Perpendicular distance from aperture to object |

| Symbol | Description |
|---|---|
| $(k_{x_t}, k_{y_t})$ | Wavenumber components for Tx aperture |
| $(k_{x_r}, k_{y_r})$ | Wavenumber components for Rx aperture |
| $(k_x, k_y, k_z)$ | Spatial frequency domain components |
| $(x_t, y_t, z_a)$ | Position of Tx in Cartesian coordinates |
| $(x_r, y_r, z_a)$ | Position of Rx in Cartesian coordinates |
| $[n_t, m_t]$ | Two-dimensional discrete position for Tx aperture |
| $[n_r, m_r]$ | Two-dimensional discrete position for Rx aperture |
| $a_t$ | Aperture of Tx array |
| $a_r$ | Aperture of Rx array |
| $a_e$ | Effective aperture |
| $a_e^{ill}$ | Illumination effective aperture |
| $\Delta a$ | Antenna spacing |
| $\mathrm{AF}_{\mathrm{Tx}}$ | Array factor of Tx aperture |
| $\mathrm{AF}_{\mathrm{Rx}}$ | Array factor of Rx aperture |
| $\mathrm{PSF}(\mathbf{r}, \mathbf{r}_v)$ | PSF value at $\mathbf{r}$ for target at $\mathbf{r}_v$ |
| $N_t$ | Total number of Tx elements |
| $N_r$ | Total number of Rx elements |
| $N_f$ | Total number of used frequencies |
| $w$ | Two-dimensional window function |
| $M$ | Measurement in actual imaging multistatic array |
| $M_m$ | Calibration measurement for match |
| $M_s$ | Calibration measurement for offset-short |
| $\Gamma_{cc}$ | Cross-coupling signal between Tx and Rx channels |
| $\Gamma_{obj}$ | Reflection from imaged object |
| $\Gamma_s$ | Reflection from offset-short calibration object |
| $\Gamma_{sim}$ | Simulation result of $\Gamma_s$ |
| $\theta_{pol}$ | Angle of mismatched polarization |
| $T$ | Combined Tx-Rx channel transfer function |
| $\mathbf{T}$ | Transfer function $T$ in matrix form |
| $\mathcal{N}$ | Gaussian noise |
| $R$ | In sec. 3.4: Distance to aperture |
| $R$ | In sec. 4.4: Sphere radius |

| Symbol | Description |
|---|---|
| $G_{ts}$ | Tx antenna gain towards specular point |
| $G_{rs}$ | Rx antenna gain towards specular point |
| $l_{ts}$ | Distance between Tx antenna and specular point |
| $l_{rs}$ | Distance between Rx antenna and specular point |

| Operator | Description |
|---|---|
| $\oint_C$ | Closed contour integration |
| $\oiint_S$ | Closed surface integration |
| $\iint_S$ | Surface integration |
| $\iiint_V$ | Volume integration |
| $\nabla(\,.\,)$ | Gradient of scalar field |
| $\nabla\cdot(\,.\,)$ | Divergence of vector field |
| $\nabla\times(\,.\,)$ | Curl of vector field |
| $\nabla^2(\,.\,)$ | Laplacian of scalar or vector field |
| $\mathcal{F}_{\text{nD}}\{.\}$ | n-dimensional Fourier transform |
| $\mathcal{F}_{\text{nD}}^{-1}\{.\}$ | n-dimensional inverse Fourier transform |
| $\Vert\,.\,\Vert$ | Length of position vector |
| $\vert\,.\,\vert$ | Magnitude of complex quantity |
| $\angle\,.$ | Phase of complex quantity |
| $\Re(\,.\,)$ | Real part of complex quantity |
| $(\,.\,)^*$ | Complex conjugate |
| $.\,*\,.$ | Convolution |
| $\hat{.}$ | In sec. 3.4: Normalization to wavelength |
| $\hat{.}$ | In sec. 4.4: Unit vector |
| $\hat{.}$ | In sec. 4.6: Statistical estimate |
| $\tilde{.}$ | In sec. 4.6: Noisy data |
| $(\,.\,)^{\text{tr}}$ | Matrix transposition |
| $\Lambda(\,.\,)$ | Matrix transposition and element-wise inversion |

| Acronym | Description |
|---|---|
| Tx | Transmitter |
| Rx | Receiver |
| SAR | Synthetic aperture radar |
| NDT | Non-destructive testing |
| SiGe | Silicon-Germanium |
| DFT | Discrete Fourier transform |
| FFT | Fast Fourier transform |
| PSF | Point spread function |
| AF | Array factor |
| PG | Processing gain |
| DBF | Digital-beamforming |
| SNR | Signal-to-noise ratio |
| RF | Radio frequency |
| RCS | Radar cross section |
| PO | Physical optics |
| GO | Geometrical optics |
| TEM | Transverse electromagnetic mode |
| PDF | Probability density function |
| AWGN | Additive white Gaussian noise |
| LO | Local oscillator |
| IF | Intermediate frequency |
| ADC | Analog-to-digital converter |
| PCI | Peripheral Component Interconnect |
| PXI | PCI eXtensions for Instrumentation |
| PXIe | PXI Express |
| DDS | Direct digital synthesizer |
| OCXO | Oven-controlled crystal oscillator |
| PCB | Printed circuit board |
| USAF | United States Air Force |
| 1D | One-dimensional |
| 2D | Two-dimensional |
| 3D | Three-dimensional |

# Bibliography

[1] M. R. Cohen and I. E. Drabkin, *A Source Book in Greek Science.* Harvard University Press, 1948.

[2] A.-H. Ibn Al-Haytham (Alhazen), *Book of Optics (Kitab Al-Manazer)*, 1021.

[3] W. Elizabeth and S. Rabah, *1001 Inventions: Muslim Heritage in Our World*, 1st ed., S. T. Al-Hassani, Ed. Manchester, UK: Foundation for Science, Technology and Civilisation, 2006.

[4] I. Newton, *Opticks or a Treatise of the Reflections, Refractions, Inflections and Colours of Light*, 1st ed. London: Printed for Sam Smith and Benj Walford. Printers to the Royal Society, at the Prince's Arms in St. Paul's Church-yard, 1704.

[5] C. Huygens, *Traité de la lumière.* Paris: Gauthier-Villars, 1690.

[6] C. Huygens and S. P. Thompson, *Treatise on Light.* University of Chicago Press (republished by project Gutenberg), 1912.

[7] T. Young, "An Account of Some Cases of the Production of Colours, not hitherto described," *Philosophical Transactions of the Royal Society of London*, vol. 92, no. 1802, pp. 387–397, 1802.

[8] J. C. Maxwell, *Treatise on Electricity and Magnetism*, 1st ed. Oxford at the Clarendon Press, 1873.

[9] K. Selvan, "A Revisiting of Scientific and Philosophical Perspectives on Maxwell's Displacement Current," *IEEE Antennas and Propagation Magazine*, vol. 51, no. 3, pp. 36–46, Jun. 2009.

[10] H. Hertz, "Ueber die Ausbreitungsgeschwindigkeit der electrodynamischen Wirkungen," *Annalen der Physik und Chemie*, vol. 270, no. 7, pp. 551–569, 1888.

[11] H. Hertz and D. E. Jones, *Electric Waves.* Dover Publications, Inc., 1962.

[12] D. Gabor, "A New Microscopic Principle," *Nature*, vol. 161, no. 4098, pp. 777–778, May 1948.

[13] D. Paganin, *Coherent X-Ray Optics.* Oxford University Press, USA, 2006.

[14] J. S. Hall, *Radar Aids to Navigation.* McGraw-Hill Co. New-York-London, M.I.T. Radiation Laboratory Series, Vol. 2, 1947.

[15] P. H. Siegel, "Terahertz Technology," *IEEE Transactions on Microwave Theory and Techniques*, vol. 50, no. 3, pp. 910–928, May 2002.

[16] H. Griffiths and N. Willis, "Klein Heidelberg – The First Modern Bistatic Radar System," *IEEE Transactions on Aerospace and Electronic Systems*, vol. 46, no. 4, pp. 1571–1588, Oct. 2010.

[17] M. Ryle, "Radio Telescopes of Large Resolving Power," *Nobel Lecture*, Jun. 1974.

[18] E. Baur, S. Buckreuß, W. Holpp, P. Honold, W. Keydel, H. Klausing, and A. Moreira, *Radar mit realer und synthetischer Apertur. Konzeption und Realisierung.*, H. Klausing and W. Holpp, Eds. Oldenbourg, 1999.

[19] D. Miller, "The TanDEM-X Satellite," in *Proceedings of the 7th European Conference on Synthetic Aperture Radar (EuSAR)*, Friedrichshafen, 2008.

[20] M. Zink, G. Krieger, H. Fiedler, I. Hajnsek, and A. Moreira, "The TanDEM-X Mission Concept," in *Proceedings of the 7th European Conference on Synthetic Aperture Radar (EuSAR)*, Friedrichshafen, Jul. 2008.

[21] G. Krieger, I. Hajnsek, K. P. Papathanassiou, M. Younis, and A. Moreira, "Interferometric Synthetic Aperture Radar (SAR) Missions Employing Formation Flying," *Proceedings of the IEEE*, vol. 98, no. 5, 2010.

[22] M. Younis, "Digital Beam-Forming for High Resolution Wide Swath Real and Synthetic Aperture Radar," Ph.D. dissertation, Universität Karlsruhe (TH), 2004.

[23] M. Soumekh, *Fourier Array Imaging.* Prentice-Hall, Inc., 1994.

[24] ——, "Bistatic Synthetic Aperture Radar Inversion with Application in Dynamic Object Imaging," *IEEE Transactions on Signal Processing*, vol. 39, no. 9, pp. 2044–2055, 1991.

[25] C. E. Cook, "Pulse Compression — Key to More Efficient Radar

Transmission," *Proceedings of the IRE*, vol. 48, no. 3, pp. 310–316, 1960.

[26] L. Kaufman and J. W. Carlson, "An Evaluation of Airport X-ray Backscatter Units based on Image Characteristics," *Journal of Transportation Security*, Nov. 2010.

[27] S. S. Ahmed, O. Ostwald, and L.-P. Schmidt, "Automatic Detection of Concealed Dielectric Objects for Personnel Imaging," in *IEEE MTT-S International Microwave Workshop on Wireless Sensing, Local Positioning, and RFID*, Cavtat, 2009, pp. 26–29.

[28] O. Ostwald and S. S. Ahmed, "Method for detecting a covered dielectric object," WO Patent Application 2010/127 739, 2009.

[29] G. R. Huguenin, "Millivision Millimeter Wave Imagers," Millimetrix, LLC, Hadley, Tech. Rep., 1997.

[30] D. M. Sheen, D. L. McMakin, and T. E. Hall, "Three-Dimensional Millimeter-Wave Imaging for Concealed Weapon Detection," *IEEE Transactions on Microwave Theory and Techniques*, vol. 49, no. 9, pp. 1581–1592, 2001.

[31] L. Yujiri, M. Shoucri, and P. Moffa, "Passive Millimeter-Wave Imaging," *IEEE Microwave Magazine*, vol. 4, no. 3, pp. 39–50, 2003.

[32] A. Luukanen, A. J. Miller, and E. N. Grossman, "Active Millimeter-Wave Video Rate Imaging with a Staring 120-Element Microbolometer Array," *SPIE Radar Sensor Technology VIII and Passive Millimeter-Wave Imaging Technology VII*, vol. 5410, 2004.

[33] D. M. Sheen, D. L. McMakin, W. M. Lechelt, and J. W. Griffin, "Circularly Polarized Millimeter-Wave Imaging for Personnel Screening," *Passive Millimeter-Wave Imaging Technology VIII, SPIE*, vol. 5789, pp. 117–126, 2005.

[34] D. L. McMakin, D. M. Sheen, J. W. Griffin, and W. M. Lechelt, "Extremely High-Frequency Holographic Radar Imaging of Personnel and Mail," *SPIE*, vol. 6201, 2006.

[35] P. Corredoura, Z. Baharav, B. Taber, and G. Lee, "Millimeter-Wave Imaging System for Personnel Screening: Scanning 10^7 Points a Second and Using No Moving Parts," *SPIE Passive Millimeter-Wave Imaging Technology IX*, vol. 6211, pp. 62 110B–1 – 62 110B–8, 2006.

[36] M. C. Kemp, "Millimetre Wave and Terahertz Technology for the Detection of Concealed Threats – a Review," *Proceedings of SPIE*, vol. 6402, pp. 64 020D–1 – 64 020D–19, 2006.

[37] E. L. Jacobs, "Concealed Weapon Identification using Terahertz Imaging Sensors," *Proceedings of SPIE*, vol. 6212, no. 2006, pp. 62 120J–1 – 62 120J–10, 2006.

[38] R. Appleby and H. B. Wallace, "Standoff Detection of Weapons and Contraband in the 100 GHz to 1 THz Region," *IEEE Transactions on Antennas and Propagation*, vol. 55, no. 11, 2007.

[39] K. B. Cooper, R. J. Dengler, N. Llombart, T. Bryllert, G. Chattopadhyay, E. Schlecht, J. Gill, C. Lee, A. Skalare, I. Mehdi, and P. H. Siegel, "Penetrating 3-D Imaging at 4- and 25-m Range Using a Submillimeter-Wave Radar," *IEEE Transactions on Microwave Theory and Techniques*, vol. 56, no. 12, pp. 2771–2778, 2008.

[40] R. Doyle, "People Screening using Millimetre-Waves — Technologies and Applications," in *International Conference on Electromagnetics in Advanced Applications*, Sep. 2009, pp. 817–818.

[41] S. Bertl, "Abbildung mit Millimeterwellen für die Personenkontrolle," Ph.D. dissertation, Technische Universität München, 2009.

[42] H. Quast and T. Löffler, "Towards Real-Time Active THz Range Imaging for Security Applications," in *International Conference on Electromagnetics in Advanced Applications*, 2009, pp. 501–504.

[43] J. N. Mait, D. A. Wikner, M. S. Mirotznik, J. van Der Gracht, G. P. Behrmann, B. L. Good, and S. A. Mathews, "94-GHz Imager With Extended Depth of Field," *IEEE Transactions on Antennas and Propagation*, vol. 57, no. 6, pp. 1713–1719, Jun. 2009.

[44] D. M. Sheen, T. E. Hall, R. H. Severtsen, D. L. McMakin, B. K. Hatchell, and P. L. J. Valdez, "Standoff Concealed Weapon Detection using a 350 GHz Radar Imaging System," in *Passive Millimeter-Wave Imaging Technology XIII*, vol. 7670, 2010, pp. 767 008–1 – 767 008–12.

[45] A. Luukanen, L. Grönberg, M. Grönholm, P. Lappalainen, M. Leivo, A. Rautiainen, A. Tamminen, J. Ala-Laurinaho, C. R. Dietlein, and E. N. Grossman, "Real-Time Passive Terahertz Imaging System for Standoff Concealed Weapons Imaging," in *Passive Millimeter-Wave Imaging Technology XIII*, vol. 7670, Apr. 2010, pp. 767 004–1 767 004–8.

[46] N. A. Salmon, R. Macpherson, A. Harvey, P. Hall, S. Hayward, P. Wilkinson, and C. Taylor, "First Video Rate Imagery from a 32-Channel 22-GHz Aperture Synthesis Passive Millimetre Wave Imager," in *SPIE Europe Security and Defence, Millimetre Wave and Terahertz Sensors and Technology*, Prague, Oct. 2011, pp.

818 805–1 818 805–12.

[47] F. Gumbmann, J. Weinzierl, P. H. Tran, and L.-P. Schmidt, “3D Millimeterwellen-Abbildung von dielektrischen Probekörpern und numerische Rekonstruktion der Materialeigenschaften,” *DGZfP-Jahrestagung*, 2007.

[48] J. B. Jackson, M. Mourou, J. F. Whitaker, S. L. Williamson, M. Menu, and G. A. Mourou, “Terahertz Imaging for Non-Destructive Evaluation of Mural Paintings,” *Optics Communications*, vol. 281, pp. 527–532, 2008.

[49] F. Gumbmann, H. P. Tran, J. Weinzierl, and L.-P. Schmidt, “Optimization of a Fast Scanning Millimetre-Wave Short Range SAR Imaging System,” in *Proceedings of the 4th European Radar Conference*, Munich, Oct. 2007, pp. 24–27.

[50] H. P. Tran, “Entwicklung und Realisierung eines echtzeitfähige Millimeterwellen-Abbildungssystems für Nahbereichsanwendungen,” Ph.D. dissertation, Universität Erlangen-Nürnberg, 2008.

[51] A. Schiessl and S. S. Ahmed, “W-Band Imaging of Explosive Substances,” in *The European Radar Conference (EuRAD)*, 2009, pp. 617–620.

[52] R. A. Hadi, S. Member, H. Sherry, J. Grzyb, Y. Zhao, W. Förster, H. M. Keller, A. Cathelin, S. Member, A. Kaiser, and U. R. Pfeiffer, “Imaging Applications in 65-nm CMOS,” *IEEE Journal of Solid-State Circuits*, vol. 47, no. 12, pp. 1–14, 2012.

[53] M. T. Ghasr, M. A. Abou-Khousa, S. Kharkovsky, R. Zoughi, and D. Pommerenke, “Portable Real-Time Microwave Camera at 24 GHz,” *IEEE Transactions on Antennas and Propagation*, vol. 60, no. 2, pp. 1114–1125, 2012.

[54] W. L. Chan, J. Deibel, and D. M. Mittleman, “Imaging with Terahertz Radiation,” *Reports on Progress in Physics*, vol. 70, no. 8, pp. 1325–1379, 2007.

[55] J. F. Federici, D. Gary, R. Barat, and Z.-H. Michalopoulou, “T-rays vs. Terrorists,” *IEEE Spectrum*, vol. 44, no. 7, pp. 47–52, 2007.

[56] A. G. Davies, A. D. Burnett, W. Fan, E. H. Linfield, and J. E. Cunningham, “Terahertz Spectroscopy of Explosives and Drugs,” *Materials Today*, vol. 11, no. 3, pp. 18–26, 2008.

[57] M. Born and E. Wolf, *Principles of Optics*, 7th ed. Cambridge University Press, 1999.

[58] E. Wolf, "Three-Dimensional Structure Determination of Semi-Transparent Objects from Holographic Data," *Optics Communications*, vol. 1, no. 4, pp. 153–156, 1969.

[59] P. Roman, *Advanced Quantum Theory.* Addison-Wesley Publishing Company, Inc., 1965.

[60] J. L. Fernandes, C. M. Rappaport, and D. M. Sheen, "Improved Reconstruction and Sensing Techniques for Personnel Screening in Three-Dimensional Cylindrical Millimeter-Wave Portal Scanning," *Passive Millimeter-Wave Imaging Technology XIV*, vol. 8022, pp. 802 205–1 – 802 205–8, 2011.

[61] T. Iwai and T. Asakura, "Speckle Reduction in Coherent Information Processing," *Proceedings of the IEEE*, vol. 84, no. 5, 1996.

[62] A. D. Wheelon, *Electromagnetic Scintillation: II. Weak Scattering.* Cambridge Univ Press, 2003, vol. 2.

[63] F. Cakoni, D. Colton, and P. Monk, *The Linear Sampling Method in Inverse Electromagnetic Scattering Theory.* SIAM, 2011.

[64] Matteo Pastorino, *Microwave Imaging*, 1st ed. John Willy & Sons, Inc., 2010.

[65] M. Soumekh, "Echo Imaging using Physical and Synthesized Arrays," *Optical Engineering*, vol. 29, no. 5, pp. 545 – 554, 1990.

[66] D. C. Munson and R. L. Visentin, "A Signal Processing View of Strip-Mapping Synthetic Aperture Radar," *IEEE Transactions on Acoustics, Speech and Signal Processing*, vol. 37, no. 12, pp. 2131–2147, 1989.

[67] J. Fortuny, "Efficient Algorithms for Three-Dimensional Near-Field Synthetic Aperture Radar Imaging," Ph.D. dissertation, University of Karlsruhe, 2001.

[68] M. Soumekh and M. Kaveh, "Image Reconstruction from Frequency Domain Data on Arbitrary Contours," in *IEEE International Conference on Acoustics, Speech, and Signal Processing*, 1984.

[69] M. Desai and W. Jenkins, "Convolution Backprojection Image Reconstruction for Spotlight Mode Synthetic Aperture Radar," *IEEE Transactions on Image Processing*, vol. 1, no. 4, pp. 505–517, 1992.

[70] L. Ulander, H. Hellsten, and G. Stenstrom, "Synthetic-Aperture Radar Processing using Fast Factorized Back-projection," *IEEE Transactions on Aerospace and Electronic Systems*, vol. 39, no. 3, pp. 760–776, Jul. 2003.

[71] J. Bredow, K. Xie, R. Porco, and M. Shah, "An Experimental Study on the Use of Multistatic Imaging for Investigating Wave-Object Interaction," *Journal of Electromagnetic Waves and Applications*, vol. 7, no. 6, pp. 811–831, 1993.

[72] G. Yates, A. Horne, A. Blake, and R. Middleton, "Bistatic SAR Image Formation," *IEE Proceedings - Radar, Sonar and Navigation*, vol. 153, no. 3, pp. 208–213, 2006.

[73] E. Hecht, *Optics*, 4th ed. Addison Wesley, 2002.

[74] S. S. Ahmed, A. Schiessl, and L.-P. Schmidt, "Multistatic mm-Wave Imaging with Planar 2D-Arrays," in *German Microwave Conference*, Munich, 2009, pp. 1–4.

[75] G. R. Lockwood and S. F. Foster, "Optimizing Sparse Two-Dimensional Transducer Arrays using an Effective Aperture Approach," in *Ultrasonics Symposium*, 1994, pp. 1497–1502.

[76] G. R. Lockwood and S. Foster, "Optimizing the Radiation Pattern of Sparse Periodic Two-Dimensional Arrays," *Science*, vol. 43, no. 1, pp. 15–19, 1996.

[77] W. L. Stutzman and G. A. Thiele, *Antenna Theory and Design*, 2nd ed. John Willy & Sons, Inc., 1998.

[78] M. Soumekh, "A System Model and Inversion for Synthetic Aperture Radar Imaging." *IEEE Transactions on Image Processing*, vol. 1, no. 1, pp. 64–76, Jan. 1992.

[79] L. N. Ridenour, "Properties of Radar Targets," in *Radar System Engineering.* McGraw-Hill Co. New-York-London, M.I.T. Radiation Laboratory Series, vol. 1, 1947, pp. 89–108.

[80] H.-J. Li, N. H. Farhat, Y. Shen, and C. L. Werner, "Image Understanding and Interpretation in Microwave Diversity Imaging," *IEEE Transactions on Antennas and Propagation*, vol. 37, no. 8, pp. 1048–1057, 1989.

[81] S. S. Ahmed, A. Schiessl, and L.-P. Schmidt, "Illumination Properties of Multistatic Planar Arrays in Near-Field Imaging Applications," in *Proceedings of the 7th European Radar Conference (EuRAD)*, Paris, 2010, pp. 29–32.

[82] A. V. Oppenheim and R. W. Schafer, *Discrete-Time Signal Processing*, 2nd ed. Prentice-Hall, Inc., 1999.

[83] The MathWorks Inc., "MATLAB," 2009.

[84] C. A. Balanis, *Advanced Engineering Electromagnetics.* John Willy

& Sons, Inc., 1989.

[85] F. A. Jenkins and H. E. White, *Fundamentals of Optics*, 3rd ed. McGraw-Hill Book Company, 1957.

[86] F. Nielsen, *Visual Computing: Geometry, Graphics, and Vision*, 1st ed. Charles River Media, 2005.

[87] F. Gustrau and A. Bahr, "W-Band Investigation of Material Parameters, SAR Distribution, and Thermal Response in Human Tissue," *IEEE Transactions on Microwave Theory and Techniques*, vol. 50, no. 10, 2002.

[88] C. Gabriel, "Compilation of the Dielectric Properties of Body Tissues at RF and Microwave Frequencies," King's College London, London, Tech. Rep., Jan. 1996.

[89] M. Zhadobov, N. Chahat, C. Le Quement, and Y. Le Drean, "Millimeter-Wave Interactions with the Human Body: State of Knowledge and Recent Advances," *International Journal of Microwave and Wireless Technologies*, vol. 3, no. 2, pp. 237–247, 2011.

[90] S. S. Ahmed, F. Gumbmann, A. Schiessl, M. Reiband, S. Methfessel, C. Maire, A. Cenanovic, O. Ostwald, C. Evers, and L.-P. Schmidt, "QPASS — Quick Personnel Automatic Safe Screening for Security Enhancement of Passengers," in *Future Security Conference*, 2011.

[91] S. S. Ahmed, A. Schiessl, and L.-P. Schmidt, "A Novel Active Real-Time Digital-Beamforming Imager for Personnel Screening," in *European Conference on Synthetic Aperture Radar (EuSAR)*, Nuremberg, 2012, pp. 178 – 181.

[92] S. S. Ahmed, A. Genghammer, A. Schiessl, and L.-P. Schmidt, "Fully Electronic E-Band Personnel Imager of 2 m^2 Aperture Based on a Multistatic Architecture," *IEEE Transactions on Microwave Theory and Techniques*, vol. 62, no. 1, pp. 651–657, 2013.

[93] D. M. Sheen, D. L. McMakin, and T. E. Hall, "Cylindrical Millimeter-Wave Imaging Technique for Concealed Weapon Detection," in *Proceedings of SPIE*, vol. 3240, 1998, p. 242.

[94] D. L. McMakin, D. M. Sheen, and H. D. Collins, "Remote Concealed Weapons and Explosive Detection on People Using Millimeter-Wave Holography," in *International Carnahan Conference on Security Technology*, 1996, pp. 19–25.

[95] S. S. Ahmed and C. Evers, "Method and device for expanding the illumination of a test object," WO Patent Application 2012/167 847, 2011.

[96] S. Trotta, B. Dehlink, R. Reuter, Y. Yin, J. John, J. Kirchgessner, D. Morgan, P. Welch, B. Knappenberger, I. To, and M. Huang, "A Multi-Channel Rx for 76.5 GHz Automotive Radar Applications with 55 dB IF Channel-to-Channel Isolation," in *European Microwave Integrated Circuits Conference*, 2009, pp. 192–195.

[97] M. Tiebout, H. Wohlmuth, H. Knapp, R. Salerno, M. Druml, J. Kaeferboeck, M. Rest, J. Wuertele, S. S. Ahmed, A. Schiessl, and R. Juenemann, "Low Power Wideband Receiver and Transmitter Chipset for mm-Wave Imaging in SiGe Bipolar Technology," in *Radio Frequency Integrated Circuits Symposium (RFIC)*. IEEE, 2011.

[98] G. Mie, "Beiträge zur Optik trüber Medien, speziell kolloidaler Metallösungen," *Annalen der Physik*, vol. 25, no. 3, pp. 377–445, 1908.

[99] P. Newman, "Contributions on the Optics of Turbid Media, Particularly Colloidal Metal Solutions," Albuquerque, 1978.

[100] R. F. Goodrich, B. A. Harrison, R. E. Kleinman, and T. B. A. Senior, "Studies in Radar Cross Sections XLVII - Diffraction and Scattering by Regular Bodies - I: The Sphere," The University of Michigan, Tech. Rep., 1961.

[101] D. B. Davidson, *Computational Electromagnetics for RF and Microwave Engineering*. Cambridge Univ Press, 2005.

[102] D. E. Kerr, *Propagation of Short Radio Waves*, 1st ed. McGraw-Hill Co. New-York-London, M.I.T. Radiation Laboratory Series, vol. 13, 1951.

[103] C. A. Balanis, R. Hartenstein, and D. DeCarlo, "Multipath Interference for In-Flight Antenna Measurements," *IEEE Transactions on Antennas and Propagation*, vol. 32, no. 1, pp. 100–104, Jan. 1984.

[104] EMSS-SA, "FEKO v5.4."

[105] E. Hecht, "Lenses," in *Optics*, 4th ed. Addison Wesley, 2002, ch. 5, pp. 175–186.

[106] G. Glaeser, "Reflections on Spheres and Cylinders of Revolution," *Journal for Geometry and Graphics*, vol. 3, no. 2, pp. 121–139, 1999.

[107] J. Schwarze, "Cubic and Quartic Roots," in *Graphics Gems*, 1st ed., A. S. Glassner, Ed. Academic Press, 1990, pp. 404–407.

[108] P. Neumann, "Reflections on Reflection in a Spherical Mirror," *American Mathematical Monthly*, vol. 105, no. 6, pp. 523–528, 1998.

[109] W. Gordon, "Reflections from Multiple Surfaces without Edges," *IEEE Transactions on Antennas and Propagation*, vol. 58, no. 10, pp. 3222–3230, 2010.

[110] C. A. Balanis, *Antenna Theory*, 3rd ed. Wiley, John & Sons, 2005.

[111] D. K. Cheng, *Field and Wave Electromagnetics*, 1st ed. Addison-Wesley Publishing Company, Inc., 1983.

[112] A. Schiessl, S. S. Ahmed, A. Genghammer, and L.-P. Schmidt, "A Technology Demonstrator for a 0.5 m x 0.5 m Fully Electronic Digital Beamforming mm-Wave Imaging System," in *European Conference on Antennas and Propagation (EuCAP)*, Rome, 2011, pp. 2606–2609.

[113] S. Methfessel and L.-P. Schmidt, "Design of a Balanced-Fed Patch-Excited Horn Antenna at Millimeter-Wave Frequencies," in *European Conference on Antennas and Propagation (EuCAP)*, Barcelona, 2010, pp. 1–4.

[114] Computer Simulation Technology AG, "Microwave Studio," 2009.

[115] R. Jünemann, C. Evers, A. Schiessl, S. S. Ahmed, and G. Hechtfischer, "Printed circuit board arrangement for millimeter wave scanners," WO Patent Application 2012/119 818, 2011.

[116] J. Bock, H. Schiffer, K. Aufinger, R. Stengl, S. Boguth, R. Schreiter, M. Rest, R. Knapp, M. Wurzer, W. Pemdl, T. Bottner, and T. Meister, "SiGe Bipolar Technology for Automotive Radar Applications," in *Proceedings of Bipolar/BiCMOS Circuits and Technology*. IEEE, 2004, pp. 84–87.

[117] A. Genghammer, "Signalverarbeitung für ein mm-Wellen Bildgebungssystem," Bachelorarbeit, Regensburg University of Applied Science, Feb 2011.

[118] S. S. Ahmed, A. Schiessl, and L.-P. Schmidt, "A Novel Fully Electronic Active Real-Time Imager Based on a Planar Multistatic Sparse Array," *IEEE Transactions on Microwave Theory and Techniques*, vol. 59, pp. 3567–3576, 2011.

[119] A. Papoulis, *Systems and Transforms With Applications in Optics.* McGraw-Hill, Inc., 1968.

[120] H. Meinke and F. Gundlach, *Taschenbuch der Hochfrequenztechnik: Band 1: Grundlagen.* Springer, 2009.

[121] S. S. Ahmed, A. Schiessl, F. Gumbmann, M. Tiebout, S. Methfessel, and L.-P. Schmidt, "Advanced Microwave Imaging," *Microwave Magazine, IEEE*, vol. 13, no. 6, pp. 26–43, Sep. 2012.

Deutsche Übersetzungen

Titel

# Elektronische Mikrowellenbildgebung mit planaren multistatischen Antennenanordnungen

# Zusammenfassung

Multistatische Antennenanordnungen zeichnen sich durch erweiterte Eigenschaften für die Mikrowellenbildgebung aus. Planare multistatische Anordnungen sind hierbei besonders gut zur praktischen Umsetzung geeignet, da sie eine Reihe von Vorzügen gegenüber den üblichen monostatischen Ansätzen bieten. Im Unterschied zu den klassischen Fernfeldverfahren erlaubt die Bildgebung im Nahbereich der Antennenanordnung eine deutliche feinere Bildauflösung und eine bessere Bildausleuchtung. So hängt beispielsweise die Tiefenauflösung in starkem Maße von der zur Bildgebung verwendeten Frequenz ab. Über die durch die Signalbandbreite gegebene Tiefenauflösung hinaus, gestattet die Nutzung der unterschiedlichen Aspektwinkel, die der multistatische Ansatz mit sich bringt, eine erhebliche Erweiterung des Raumfrequenzbereichs und eine daraus resultierende Steigerung der Tiefenauflösung. Anders als bei konventionellen Entfernungsmessradaren ermöglichen planare multistatische Antennenanordnungen eine Tiefenfokussierung selbst bei der Signalbandbreite Null. Hierbei kommt der Höhe der Frequenz selbst und nicht nur dem Frequenzhub eine wichtige Bedeutung bei der Erreichung einer qualitativ hochwertigen Fokussierung zu. Darüber hinaus hängen auch die Phaseninformationen der rekonstruierten Mikrowellenbilder vom verwendeten Frequenzbereich ab. Die Phase ist speziell nützlich, um Eigenschaften der abgebildeten Objekte weit unterhalb der Wellenlänge erkennen zu können. Die Bildgebung im Hochfrequenzbereich, insbesondere im Millimeter- und Sub-Millimeterwellenlängenbereich kann daher zu Recht als ein vorteilhaftes Verfahren zu Erzeugung von dreidimensionalen Mikrowellenbildern von herausragender Güte angesehen werden.

Voll besetzte oder auch nur dicht besetzte Antennenanordnungen lassen sich wegen ihrer Komplexität praktisch kaum verwirklichen. Der Schlüssel zur technischen Realisierbarkeit liegt in einem geeigneten Konzept zur Ausdünnung der Antennenanordnung, dessen Ziel es sein muss, einerseits sowohl die Komplexität der Hardware als auch die Messdatenaufnahmezeit, die Leistungsaufnahme und die Herstellkosten zu minimieren und andererseits hierbei die angesprochene hohe Bildqualität möglichst nicht zu verlieren. Die Synthese einer geeignet ausgedünnten Antennenanordnung, die noch dazu hinreichend einfach fertigbar sein muss, ist keine triviale

Aufgabe, sondern stellt eine echte Herausforderung dar, die nur auf der Basis eines umfassenden Verständnisses der Wirkungsweise von planaren multistatischen Antennenanordnungen gelöst werden kann. Mithilfe optimiert ausgedünnter Antennenanordnungen konnte die Erfüllbarkeit dieser Anforderungen bewiesen werden, wobei gleichzeitig eine erhebliche Verminderung der Gesamtanzahl der Antennen erzielt wurde. Eine Anordnung mit einer Aperturfläche von einem Quadratmeter im Frequenzband 70 GHz bis 80 GHz wurde synthetisiert. Hierbei kommen insgesamt 1536 Sendantennen und ebenso viele Empfangsantennen zum Einsatz, von denen jeweils 96 auf sechzehn ebenfalls quadratischen identischen Antennengruppen verteilt werden. Dadurch erreicht man eine Verminderung der Antennenanzahl um den Faktor 147 im Vergleich zu einer voll besetzten monostatischen Anordnung mit denselben Abmessungen. Die experimentelle Überprüfung bewies die gute Übereinstimmung der tatsächlichen Eigenschaften des Systems mit den vorhergesagten Werten. Besonders wertvoll aus fertigungstechnischer Sicht ist darüber hinaus das erarbeitete modulare Antennenkonzept, dass eine volle Skalierbarkeit und, daraus resultierend, zugeschnittene flexible Lösungen für eine Palette von unterschiedlichen, herausfordernden Anwendungsmöglichkeiten eröffnet.

Bei der Bildgebung von Objekten mit im Vergleich zur Wellenlänge glatten Oberflächen sind besondere Ausleuchtungscharakteristika zu beachten, die durch spekulare Reflexionen hervorgerufen werden. Der allgemeine Zusammenhang zwischen der Geomtrie der Antennenanordnung und der daraus resultierenden Ausleuchtung wurde umfassend betrachtet. Durch Redundanz in der effektiven Apertur planarer multistatischer Antennenanordnungen wird eine signifikante Verschlechterung der Homogenität der Ausleuchtung im fokussierten Bild verursacht. Um diese Nachteile zu überwinden, wurde ein geeignetes Gewichtungsverfahren entwickelt, das die Ausleuchtungsinhomogenitäten kompensiert und eine gleichmäßige Bildausleuchtung ergibt. Darüber hinaus verhalten sich innerhalb des Fokussierprozesses die spekularen Reflexionen anders als die diffusen, wodurch Phasenabweichungen verursacht werden, die die maximal erreichbare Bildhelligkeit begrenzen. Hierzu wurde eine quantitative Analyse durchgeführt mit dem Ziel, die theoretischen Grenzen der für multistatische Antennenanordnungen erreichbaren Bildhelligkeit spezifizieren zu können, und zwar unter Beachtung sowohl spekularer als auch diffuser Streumechanismen. Schließlich wurde eine geeignete numerische Simulationsmethode entwickelt, die eine hinreichend präzise Vorhersage der erreichbaren Ausleuchtungsgüte für im Vergleich zur Wellenlänge große und beliebig kompliziert geformte dreidimensionale Oberflächen ermöglicht. Diese Simulationsmethode wurde experimentell überprüft und anschließend verwendet, um die Ausleuchtung menschlicher Körper für eine typische Personen-Scanner-Anwendung zu ermitteln.

Für eine qualitativ hochwertige Bildgebung im Mikrowellen-Frequenzbereich erweist sich eine geeignete Systemkalibrierung als unverzichtbar. Die dadurch erreichbare Etablierung einer wohldefinierten Referenzphasenfläche ist die entscheidend notwendige Grundlage für eine synthetische Fokussierung mit ausreichender Genauigkeit. Zur praktischen Inbetriebnahme des entwickelten Bildgebungsverfahrens muss eine handliche und effiziente Kalibriermethode zur Anwendung kommen. Hierzu ist ein Kalibriermessobjekt mit genau bekanntem komplexen Reflexionsfaktor nötig. Aus praktischen Gründen wird der Einsatz von im Vergleich zur Wellenlänge großen reflektierenden Kugeln als geeignet angesehen. Kurioserweise ist bisher keinerlei analytische Lösung für das Rückstreuverhalten von Kugeln im Nahfeld bistatischer Antennenanordnungen veröffentlicht worden. Deshalb wurde im Rahmen der vorliegenden Arbeit ein geeigneter elektromagnetischer Simulator entwickelt und erfolgreich verifiziert. Außer der Systemfehlerkorrektur für das bildgebende System wurde ein Verfahren zur Geometriefehlerkorrektur erarbeitet und untersucht. Bei der Bildgebung im Wellenlängenbereich von Millimetern oder weniger wird eine entsprechend genaue Positionierung und Ausrichtung relativ großer Kalibriermessobjekte praktisch immer schwieriger. Daraus resultieren Unwägbarkeiten der Kalibrierung, die die Bildqualität drastisch verschlechtern können. Dies wird durch die entwickelte Geometriefehlerkorrektur verhindert, die sowohl analytisch als auch numerisch untersucht wurde. Außer den erwähnten System- und Geometriefehlern stellt das Signalrauschen ein weiteres unvermeidbares Problem dar. Dank des multistatischen Systemansatzes lässt sich jedoch der Rauscheinfluss auf die Kalibriermesswerte erheblich reduzieren. Hierzu dient eine Signalkorrelation auf Basis von effizienten Matrizenoperationen, womit eine deutliche Verbesserung des Signal-Rausch-Abstandes erreicht werden konnte.

Zur Verifikation der erzielten Ergebnisse wurde ein vollelektronischer Demonstrator entworfen und aufgebaut. Er besteht aus einer Antennenanordnung von jeweils 736 Sende- und Empfangsantennen auf einer Grundfläche von 50 cm mal 50 cm. Die Geometrie der Antennenanordnung entspricht der vorher erwähnten Topologie, allerdings im verkleinerten Maßstab, womit zugleich die tatsächliche Skalierbarkeit des Konzepts bewiesen wurde. Der Demonstrator arbeitet im Frequenzband von 68 GHz bis 82 GHz mit einem schrittweise in der Frequenz springenden Dauerstrichsignal. Die Systemarchitektur fußt auf einem digitalen Strahlformungskonzept, bei dem die Bildfokussierung rein synthetisch unter Benutzung eines industriellen Rechners erzeugt wurde. Eine hohe Bildqualität mit einer lateralen Auflösung von 2,2 mm bei einer Bilddynamik von mehr als 30 dB konnte erzielt und nachgewiesen werden. Dank der schnellen Messdatenaufnahme wurde die Erzeugung von Mikrowellen-Bildern ohne Bewegungsartefakte auch von Menschen erfolgreich durchgeführt.

Insgesamt verdeutlichen die erreichten theoretischen und praktischen Ergebnisse klar die Vorzüge planarer multistatischer Antennenanordnungen. Nichtsdestoweniger zeigt sich die technologische Umsetzung des Konzepts als sehr herausfordernd, denn sie verlangt eine hohe Integration sowohl im analogen HF-Frontend als auch im digitalen Backend. Die aufwändige Algorithmik zur Bildrekonstruktion stellt den entscheidenden Flaschenhals dar, der Bildwiederholfrequenzen für Echtzeit-Videoraten zur Zeit noch verhindert. Man kann jedoch zuversichtlich sein, dass der zu beobachtende stetige Fortschritt auf dem Gebiet der digitalen Rechenwerke ausreichen wird, um auch zukünftige Anforderungen lösen zu können. Darüber hinaus kann der aufgenommene Datenschatz von komplexwertigen, dreidimensionalen, hochaufgelösten Messwerten mit hoher Signaldynamik auch für weiter fortgeschrittene Verfahren zu Super-Auflösung und weiter verbesserte Signal- und Bildverarbeitungsalgorithmen genutzt werden, um weitere Eigenschaften und Spezifika der mithilfe des bildgebenden Verfahrens untersuchten Objekte zu ermitteln. Auf diese Weise wird ein weites Feld von neuen Applikationen addressiert, die diese moderne Form der Mikrowellenbildgebung effizient nutzen können.